THEORY OF SATELLITE
AND
MOBILE (CELLULAR) TELECOMMUNICATIONS

THEORY OF SATELLITE
AND
MOBILE (CELLULAR) TELECOMMUNICATIONS

Transmission Systems Planning, Design and Analysis

Ashok K. Sinha

Copyright © 2015 by Ashok K. Sinha.

ISBN: Softcover   978-1-5035-6659-0

All rights reserved. No part of this book may be reproduced or transmitted in any form or by any means, electronic or mechanical, including photocopying, recording, or by any information storage and retrieval system, without permission in writing from the copyright owner.

Any people depicted in stock imagery provided by Thinkstock are models, and such images are being used for illustrative purposes only.
Certain stock imagery © Thinkstock.

Print information available on the last page.

Rev. date: 07/10/2015

To order additional copies of this book, contact:
Xlibris
1-888-795-4274
www.Xlibris.com
Orders@Xlibris.com
549647

CONTENTS

Part I: SATELLITE TELECOMMUNICATIONS 1

1. SATELLITE LAUNCH 3

1.1. Introduction 3
1.2. The Rocket Engine 5
1.3. The Rocket Equation 6
1.4. Specific Impulse 9
1.5. Elements of Orbital Analysis 12

    1.5.1. Equation of Motion under Newton's Law of Gravitation 12

1.6. Basic Types of Satellite Orbits and Applications 23

    1.6.1. Geostationary Satellites (GEO) 23
    1.6.2. Non-Geostationary Satellites 24
    1.6.3. Sun-Interference 30

2. MAJOR SYSTEM COMPONENTS (SUBSYSTEMS) 34

2.1. Satellite (Space Segment) 34

    2.1.1. Repeaters or Transponders 35

        2.1.1.1. Traveling Wave Tube Amplifier (TWTA) 36
        2.1.1.2. Solid Sate Power Amplifier (SSPA) 37

    2.1.2. Solar Panels 38
    2.1.3. Battery System 39

2.1.4. Attitude Control Subsystem (ACS) 39
2.1.5. Satellite Antennas 40
2.1.6. Thermal Control Subsystem (TCS) 41

2.2. Earth Stations (Earth Segment) 41

2.2.1. Earth Station Antenna 42
2.2.2. High Power Amplifier (HPA) 43
2.2.3. Low Noise Amplifier (LNA) 44

3. THE LINK EQUATION 45

3.1. General 45
3.2. Link-Budget Formulation 47

3.2.1. Uplink 47
3.2.2. Downlink 51
3.3.1. Total Link 52
3.3.2. Software Implementation and Flow-Chart 52

3.3. Atmospheric Propagation 55

4. DIGITAL TECHNOLOGY 59

4.1. General 59
4.2. Signal Modulation 60

4.2.1. Binary Phase Shift Keying (BPSK) 60
4.2.2. Quaternary Phase Shift Keying (QPSK) 60
4.2.3. Quadrature Amplitude Modulation (16-QAM) 62

4.3. Signal Detection 63
4.4. Error Correction 63
4.5. Baseband Signal 67

4.5.1. Digital Voice (Telephony) 69

4.5.1.1. Voice Coding by Source Coding 70

  4.5.1.2. Waveform Coding 71
  4.5.1.3. PCM 71

5. MULTIPLE ACCESS TECHNIQUES 76

 5.1. Introduction 76
 5.2. Frequency Division Multiple Access (FDMA) 76
 5.3. Time Division Multiple Access (TDMA) 78
 5.4. Code Division Multiple Access (CDMA) 82

  5.4.1. Introduction 82
  5.4.2. Spread-Spectrum CDMA (SS-CDMA) 84
  5.4.3. Frequency Hopping CDMA (FH-CDMA) 89

6. CODING – I 92

 6.1. Introduction 92
 6.2. Basic of Information Theory 93
 6.3. Shanon's Theorem 95
 6.4. Bandwidth-Power Tradeoff 96
 6.5. The Effect of White Noise 97
 6.6. Modular Arithmetic 100
 6.7. Shift Register 100
 6.8. Hamming Distance 101
 6.9. Algebraic Coding Theory 102
 6.10. Polynomial Coding 107
 6.11. Primitive Roots 113
 6.12. Matrix Representation 115
 6.13. Syndrome 118
 6.14. Coding Efficiency 120
 6.15. Coding in Satellite Communications 122

7. CODING – II 125

 7.1. Types of Coding 125
 7.2. Block Codes 125
 7.3. Cyclic Codes 130
 7.4. Reed Solomon Code (RS Code) 133

7.5. Turbo Code 134
7.6. Low Density Parity Check (LDPC) Codes 135
7.7. Convolution Coding 137
7.7. Viterbi Decoding 139

8. FREQUENCY BAND ALLOCATION: INTER-SYSTEM INTERFERENCE (II) 141
9. CELLULAR (MOBILE) COMMUNICATION SYSTEMS 145

9.1. Introduction 145
9.2. The B2G Systems 146
9.3. The B3G System 148

    9.3.1. General 148
    9.3.2. Frequency Bands 150
    9.3.3. IEEE 802.11 Systems 150
    9.3.4. IEEE 802.11 Technologies 152

        9.3.4.1. Infrared (IR) Technology 152
        9.3.4.2. Direct Sequence Spread Spectrum (DSSS) 153
        9.3.4.3. Frequency Hopping Spread Signal (FHSS) 153
        9.3.4.4. Orthogonal Frequency Division Multiplexing (OFDM) 154
        9.3.4.5. Orthogonal Frequency Division Multiple Access (OFDMA) 155
        9.3.4.6. IEEE 802.16-2004 155
        9.3.4.7. Wireless-Fidelity (Wi-Fi) 158
        9.3.4.8. Hot Spot30 159
        9.3.4.9. 3G Networks 159

9.4. The 4G System 163

10. TYPES OF SIGNALS 168

10.1. Introduction 168
10.2. Digital Voice 169

10.2.1. General 169
10.2.2. Signal-to-Noise Ratio (S/N) for
Quantization Noise 171

    10.2.2.1. Signal to Quantization Noise Ratio 171
    10.2.2.2. The µ-Law (North American Standard) ... 173
    10.2.2.3. The A-Law (European Standard) 173

10.2.3. Signal-to-Noise Ratio (S/N) for Thermal
Noise 174
10.2.4. Phase Shift Keying (PSK) 177

    10.2.4.1. Binary PSK (BPSK) 177
    10.2.4.2. Offset Quarternary PSK (OQPSK) 178
    10.2.4.3. Eight-Phase PSK (8ϕPSK) 179

**10.3. Bit Error Rate (BER)
for Phase Shift Keying (PSK) Modulation** 182

    10.3.1. BPSK 182
    10.3.2. QPSK 183
    10.3.3. 8-Phase PSK (8ϕPSK) 184

10.4. Digital Hierarchy 185
**10.5. Multiple Access and
Related Problems and Performance** 186

    10.5.1. Introduction 186
    10.5.2. Frequency Division Multiple Access
    (FDMA): Intermodulation 187
    10.5.3. Time Division Multiple Access (TDMA) 189
    10.5.4. Code Division Multiple Access (CDMA) in
    Mobile Channels 192

        10.5.4.1. General 192
        10.5.4.2. CDMA Output Power 194
        10.5.4.3. Modulation of CDMA Traffic Signal 195
        10.5.4.4. Transmission Sequence 196

10.5.4.5. Call Processing 198

10.6. Digital Television 199

10.6.1. General 199
10.6.2. Typical Video Parameters 200
10.6.3. Video Color 201
10.6.4. High Definition Television (HDTV) 201

11. SPECIAL TOPICS IN MOBILE SERVICE SYSTEMS 203

11.1. Satellite Link Design 203
11.2. Satellite Antenna Beam Coverage Pattern 206
11.3. Cellular Pattern of Multiple Spot Beams 210
11.4. Doppler Effect and Range-Rate 215
11.5. Variation of the Satellite Apparent Position 218
11.6. Change in the Period of Revolution 219
11.7. Mobile Systems Beam/Cell Coverage 222

11.7.1. Introduction 222
11.7.2. Ground Reflection of the Incident Wave 223
11.7.3. System Efficiency 231
11.7.4. CDMA Systems Design 236
11.7.5. Antenna Coverage Pattern (ACP) for
Cellular/Mobile (CEMO) Systems 244
11.7.6. Spectral Density for Mobile-to-Mobile
Communication 248

11.7.6.1. Introduction 248
11.7.6.2. The Transfer Function 248
11.7.6.3. Spectral Distribution 251
11.7.6.4. The Link Equation for a Mobile
Receiver 254

**12. EPILOG: SATELLITE
TELECOMMUNICATIONS AND BASIC PHYSICS 259**

PROLOGUE

It is not very common to see a book on a subject unambiguously categorized as engineering be authored by someone unambiguously *not* categorized as an engineer, professionally.

Academically and professionally, I call myself a theoretical physicist, primarily interested in the specific branch of physics called *theoretical elementary particle physics* (or *theoretical High Energy Physics*) and *cosmology* – a special branch of physics which strives to enhance the understanding of the basic building blocks or constituents of matter on the one hand and the origin and evolution of the universe, on the other. The first of these fields is the science of the extremely microscopic systems most commonly described by quantum field theory; and the second, dealing with extremely large systems of stars and galaxies commonly described by general relativity theory of Einstein. Indeed, the author of this book did also write a book – called "New Dimensions in Elementary Particle Physics and Cosmology" – based on the results of his own independent theoretical results on the subjects this title encompasses, with a simple mathematical model for elementary particle masses including the masses of the so-called Higgs boson, popularly also known as "The God Particle" (Please see the List Of Books preceding this Prologue).

In any case, this book in your hand ("Theory of Satellite and Mobile Telecommunications") is mainly focused on *only* the theoretical aspects of some of the basic areas in connection with the planning, design, and analysis of a satellite and mobile telecommunications systems, without any attempt at all to provide any material necessary that goes into many practical aspects of the subject (equipment and hardware design and construction, system implementation, system operation, cost model, technical or performance-vs-cost trade-off analyses, and so on.) Practicing engineers interested in these latter (practical) aspects could leaf through this book only for certain *theoretical* backgrounds and mathematical equations and formulae involved, and should be better advised to consult one or more of several other excellent books and papers providing such engineering details (see References provided in this book.)

I thought it is necessary to unconditionally clarify this important point before the beginning of this book. It is true that, in course of his career, the author immensely benefitted from his professional association with world-famous organizations like COMSAT (Communications Satellite Corporation of USA) and INTELSAT [International Telecommunications Satellite Organization, the international cooperative that is the main provider of satellite telecommunications (SATCOM) throughout the world with membership of more than 200 countries], the vanguard institutes for research and development s and implementation of this unique technology. Many significant contributions were made by him in the research and development phase of this field, with publication of numerous research papers in professional journals, internal communications, and participation and organization of a great many international conferences during early 1970s through late 1990s, and subsequent independent consulting work. Both of the mentioned organizations are located in Washington D.C. in USA (COMSAT has been dissolved now, and INTELSAT now also has a branch in Europe.)

In particular, the author played an active role in research and development of SATCOM for relatively small and isolated users (e.g., the small island countries of the South Pacific region, oil rigs, Very Small Aperture Terminal or VSAT Users, rural telecommunications, etc.); optical inter-satellite links (ISL), and digital satellite television broadcast systems. In many of these areas, the author had the good opportunities of representing INTELSAT in the International Consultative Committee for Radio (CCIR); International Consultative Committee for Telephone and Telegraph (CCITT) in Geneva, Switzerland; and in international conferences and seminars throughout the world (Thailand; China; Japan; Hungry; France; Denmark; Canada; Colombia, South America; Mali, Africa; New Zeeland, Australia; Hawaii, USA; U.N.O., New York; etc.) Furthermore, he also actively participated in the planning and design of domestic and regional satellite systems for several countries and organizations including NASA, ARABSAT, CHINASAT, AUSTRALIASAT, COLOMBIASAT, etc. For such widespread and intensive involvement and active contributions in a field of engineering, *on part of a theoretical particle physicist,* may not be common; but I thoroughly enjoyed these professional activities, as much as working on intricate problems in the forefront of theoretical High Energy Physics and Cosmology. I regard such a combination as an especial fortune in terms of playing a double role in the two fields. This book is a link in the chain of same type of duality.

In this book, ample references have been made of a few selected sources, including reproduction of certain Illustrative Figures and Drawings as well as analyses and formulae. It is my pleasure to acknowledge my indebtedness of the following pathfinder authors (and their books) in particular: William C.Y. Lee ("Wireless & Cellular Telecommunications," Third Edition); M. Richharia ("Satellite Communications Systems," Second Edition); Richard W. Hamming ("Coding and Information Theory"); and, last but not the least, Wilber L. Pritchard et al ("Satellite Communication Systems Engineering.") I am proud to mention, also, that I had the wonderful opportunity to work with Late Dr. Wilber Pritchard, the first Director of COMSAT Laboratories in Clarksburg, MD, where I worked when I started my first ever introduction with the field of satellite telecommunications, as part of the development of the newly introduced TDMA (Time Division Multiple Access) technology. As a part of the Consulting Group later started and headed by Dr. Pritchard, in Bethesda, MD, I greatly and gratefully enjoyed my task and sole responsibility of developing the optimal specifications of the electromagnetic or transmission subsystem of ECHOSTAR-1, the first digital TV transmission and broadcast satellite system in the world (started by Charles Ergen) – now providing a multitude of Digital TV services including Satellite News Gathering (SNG) under the commercial name of "DISH NETWORK," throughout the world. It must also be mentioned that any error of omission and commission in this book is solely my responsibility, and not in the least of any of the above referred authors and sources.

It is my genuine pleasure to express my gratitude to a number of pioneers of satellite telecommunications: Jack Dicks, Dr. Joseph Pelton, Late Dr. Burton Edelson, Late Marcel Perras, Dr. Willian

Wu, Late Dr. Osamu Shimbo, Dr. Peter Nuspl, Dr. Ashok Kaul, Dr. Prakash Chitre, Dr. M. Muratani, to name just a few. When satellite communications has become a household word and a routine provider of all types of services globally, a remembrance of these early stalwarts in this most interesting modern field and technology remains fresh. Applications of satellite telecommunications for the INERNET and for mobile communications are proliferating by the day. It is well to recall that all the projects and activities in space explorations – be it Apollo Moon Mission of NASA, the PLANCK Satellite Mission of EUROSAT, or the MOM (MARS ORBITAL MISSION) MANGALYAN program of India – would be impossible without the use of the state-of-the-art applications of satellite telecommunications technology. New scientific and technological innovations – some illustrative examples of which are presented in the EPILOGUE of this book -- are bound to keep this field in the forefront of finest human endeavors for days and years yet to come.

■ Ashok Sinha   Santa Clara, California
                                 February 25, 2015.

Part I

SATELLITE TELECOMMUNICATIONS

1. SATELLITE LAUNCH

1.1. Introduction

Provision of telecommunications across the world with the help man-made satellites orbiting the earth is now a common and familiar phenomenon. Satellites are launched into space using launch vehicles which work on the principle of rocket action. The rocket or launch vehicle houses one or more satellite(s), referred to as the payload, during the launch process. The trajectory of the rocket is monitored and controlled by ground commands, and finally the payload is delivered at appropriate orbital position. The satellite is stabilized and its orbital location is also monitored and controlled by ground command throughout the life of the satellite.

A satellite could be launched into the so-called geostationary orbit in the earth's equatorial plane. In this circular orbit, the rotational speed of the satellite in its orbit becomes equal to the rate of diurnal rotation of the earth about its axis (i.e., about 24 hours); so the satellite appears stationary with respect to an antenna on the earth's surface (*earth station*) accurately pointed to the satellite (*space station*). The altitude of the geostationary orbit above the equator is about 23,500 miles (35,800 km). Satellites in an orbit below the geostationary orbit would rotate faster than the earth's rotational motion, while those in a higher orbit than the geostationary orbit would rotate at a speed lower than the earth's rotational speed.

A set of three geostationary satellites spaced 120° apart in longitude with respect to one another can provide a full and continuous coverage of the entire earth's surface (except the North and South Polar regions). This type of satellite system could therefore be used to provide communication between any two points on the earth, making global telecommunication via (geostationary) satellites possible. This fact was first pointed out by famed British scientist and science-fiction writer, Arthur C. Clarke, as early as in 1945 [Clarke, 1945]. The first successful geostationary satellite, SYNCOM- III, was launched by NASA in 1964. The first commercial geostationary telecommunications satellite was launched by the International Telecommunications Satellite Organization (INTELSAT) in April 1965. This satellite, called INTEL SAT- I or *Early Bird*, was placed over the Atlantic Ocean to provide communication links between the East Coast of USA and Western Europe.

At the present time, over 300 geostationary satellites providing international, regional, and domestic telecommunications to more than 200 countries around the globe are operational. The longitudinal positions, frequency band, antenna coverage patterns, and other factors are carefully selected to avoid or minimize electromagnetic interference between adjacent satellites in the geostationary orbit.

The satellites are launched into the desired orbit using powerful rockets. The launch operation is mathematically describable by the so-called *rocket equation* based on the principle of conservation of linear momentum which, in turn, relates to Newton's Third Law of Motion: *"Corresponding to every action, there is an equal and opposite reaction."*

In the following section, the basic elements of the rocket engine are indicated. Then the *rocket equation* is presented, and further analyzed to describe various essential operational steps and important parameters involved in the launch process.

- Clarke, Arthur C., "Extra-Terrestrial Relays," Wireless World, October 1945.

1.2. The Rocket Engine

The basic components of a general rocket engine are illustrated in Figure 1.1:

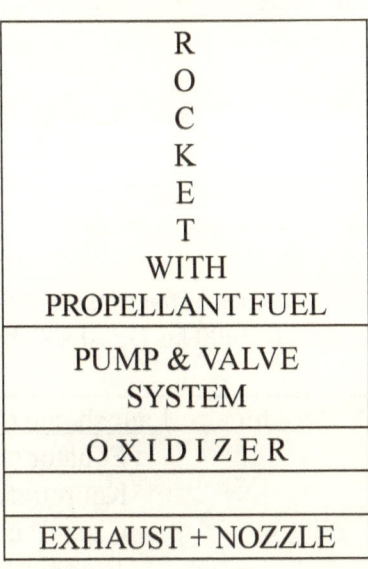

Figure 1.1. Schematic Diagram of a Rocket Engine and Payload

As shown in Figure 1.1, a suitable propellant is pumped into the combustion chamber where it is ignited to produce exhaust gases at a very high pressure and velocity, in a properly controlled manner. This action, and the resulting downward momentum of the exhaust gases causes an equal and opposite reaction *(Newton's Third Law of Motion)*, making the rocket propel upward with equal momentum *(Conservation of Momentum)*. In course of its flight, the rocket engine's fuel is expended. The motion of the rocket is controlled by ground command as necessary.

The launched system is first stabilized in a circular orbit near the earth, called the *Parking Orbit*. As a second stage, the orbit is made elliptical with perigee equal to the radius of the Parking Orbit and apogee equal to the geostationary altitude. This intermediate configuration is called the *Transfer Orbit*. As a final stage of the launch process, the system is boosted into a circular, geostationary orbit, and gradually drifted into the desired longitudinal position. Various phases of the launch process are carefully monitored and maneuvered by means of *Telemetry, Tracking and Control (TT&C)* system on ground.

The *Beginning of Life (BOL)* mass of the satellite system depends on the Transfer Orbit mass and the launch site location (Latitude). For example, for 1000 kg transfer orbit mass, the beginning-of-life (BOL) mass of the satellite is approximately as shown in Table 1.1.

Table 1.1 – Examples of the BOL Satellite Mass Corresponding to 1000 kg Transfer Orbit Mass[1]

BOL Mass (kg)	for	Launch-site (Latitude, degrees)
615		Equator (sea launch, 0°)
612		Kourou (5.23°)
547		Cape Canaveral (28.5°)
473		Tyuratam (46°)

Lower latitude launch sites imply less fuel to correct the inclination, hence higher BOL satellite mass.

1.3. The Rocket Equation

The motion of the rocket can be described mathematically in a simple manner, with respect to a suitably chosen inertial (coordinate) system. Let M and V be the instantaneous (initial) mass and velocity, respectively, of the propellant plus payload (satellite) in the upward (or forward) direction. In a short interval of time, dt, the amount of the propellant expended and exiting the exhaust chamber in the downward (or backward) direction is assumed to be *dm*, and during this time the rocket (propellant plus payload) velocity is assumed to increase from v to $v + dv$, its mass reducing from m to $(m-dm)$. Assuming the velocity of the exhaust, relative to the rocket, to be v_e, we must have, by conservation of momentum along the direction of the flight, the equality in the inertial frame):

$$mv = (m-dm)(v+dv) + dm(v + v_e) \qquad (1.1a)$$

Neglecting the second order term (dmdv), we get:

$$\frac{dv}{dm} = -\frac{v_e}{m} \qquad (1.1b)$$

If v_e remains constant, we have, by integrating the above equation:

$$\int_{v_0}^{v} dv = -v_e \int_{m_0}^{m} \frac{dm}{m}$$

where m_o is the initial mass of the propellant (plus the payload) and v_o is the initial velocity. Thus we have the result:

$$\boxed{v - v_o = v_e \ln\left(\frac{m_o}{m}\right)} \qquad (1.2a)$$

Equation (1.2a) is the desired **Rocket Equation** which can also be written in the form:

$$\Delta v = v - v_o = v_e \ln\mu \qquad (1.2b)$$

where $\mu = \frac{m_o}{m}$ is called the *mass ratio*, and Δv represents the net increase in the velocity of the rocket. When all the expendable propellant has been exhausted, the corresponding final mass of the rocket (plus the payload) becomes m. The expendable propellant mass is:

$$\Delta m = m_o - m$$

so that the rocket equation could also be written as:

$$\Delta v = -v_e \ln \frac{m}{m_o} = -v_e \ln \frac{m_o - \Delta m}{m_o}$$

i.e.,

$$\Delta v = -v_e \ln \left(1 - \frac{\Delta m}{m_o}\right) \qquad (1.2c)$$

Or,

$$\Delta v = -v_e \ln (1 - \xi) \qquad (1.2d)$$

where:

$$\xi = \frac{\Delta m}{m_o} = \frac{m_o - m}{m_o} = 1 - \frac{1}{\mu} \qquad (1.2e)$$

is called the *propellant mass fraction*. The propellant mass required to achieve a given increase Δv in the rocket velocity is given from Equation (1.2c) as:

$$\Delta m = m_o [1 - \exp(-\frac{\Delta v}{v_e})] \qquad (1.2f)$$

The choice of the type of the propellant (e.g., ion propulsion, chemical mixtures, solid state propellant, photon engines, etc.) usually determines the *relative velocity*, v_e, and hence the efficiency of the rocket action and operation.

Equation (1.1b) can also be written in a form expressing the rate of change of velocity (i.e., the acceleration of the rocket) as related to the rate of change of the rocket (propellant plus payload) mass, with respect to time:

$$m \frac{dv}{dt} = - v_e \frac{dm}{dt} = M_t \qquad (1.2g)$$

where $M_t = - v_e \frac{dm}{dt}$, called the *momentum thrust*, is a positive quantity (since $\frac{dm}{dt} < 0$), and is in units of force.

1.4. Specific Impulse

It is useful to introduce the so-called *effective exhaust velocity*, v_{eff}, defined as:

$$v_{eff} = \propto v_e = \frac{(p_e - p_a) A_e}{dm/dt} \qquad (1.3a)$$

where $\propto$ *is* a correction factor depending on the size and shape of the exhaust nozzle, ($p_e - p_a$) and represents the pressure mismatch and non-axial flow of the propellant, and A_e is the exhaust nozzle exit area. Clearly, for $\propto = 1$ and $p_e = p_a$, $v_{eff} = v_e$. Note that since $\frac{dm}{dt} < 0$, the second term on the right is positive ($p_e < p_a$).

The **specific impulse** of a rocket system is defined as:

$$I_{sp} = \frac{M_t}{\left[\frac{dm}{dt}\right] g_o} \qquad (1.3b)$$

where the brackets [] in the denominator on the right denote the absolute magnitude (without the negative sign of $\frac{dm}{dt}$), and g_o is a *"reference" acceleration due to gravity* of a hypothetical location, with the standard value:

$$g_o = 980.665 \text{ cm/sec}^2.$$

Thus the specific impulse is a ratio of units of force to the rate of change of the propellant weight (and not mass) with respect to the above standard value of the acceleration due to gravity at the hypothetical reference location (and not at the local acceleration due to gravity, g). The unit of specific impulse in seconds is:

$$I_{sp} = \frac{v_{eff}}{g} \qquad (1.3c)$$

The propellant mass required to produce a given amount of change in the velocity can be written in terms of the specific impulse by using Equation (1.2f) and Equation (1.3c) as:

$$\Delta m = m_o - m = m_o \left[1 - \exp\left(-\frac{\Delta v}{I_{sp} g_o}\right)\right] \qquad (1.4a)$$

The values of the specific impulse for a few propellant types are given below (Table 1.2). The specific impulse values specified are for vacuum; at sea level, the actual values are somewhat (typically about 20%) less.

Table 1.2
Typical Parameter Values of Certain Propellants[1]

Fuel*	Oxidizer*	Density (gm/cc)	Specific Impulses
Hydrogen (H_2)[C]	Oxygen (O_2) [C] {L}	0.326	430
Hydrogen (H_2)[C]	Fluorine (F_2) [C] {L}	0.539	440
Hydrazine (N_2H_4) [N]	Chlorine Trifluoride (ClF_3)[C]{L}	1.501	312
Double base	Nitroglycerin {S} (1.4%)	1.66	310
Polyvinyl chloride	Nitroglycerin {S} (74%)	1.63	266
Polybutadiene	Nitroglycerin {S} (70%)	1.77	289

- The following abbreviations have been used in this Table:

    [C]: cryogenic ; [N]: non-cryogenic (storable)

    {L}: liquid propellant ; {S}: solid propellant (Oxidizer)

Note: For a large number (> 60) of different combinations of Fuels and Propellants Oxidizers, see Reference [1].

1.5. Elements of Orbital Analysis

We first derive a general solution of a two-body problem subject to the universal gravitational attraction (Newton's Law of Universal Gravitation), and then consider special cases of interest for our purpose.

1.5.1. Equation of Motion under Newton's Law of Gravitation

Using the coordinates shown in Figure 1.2, Newton's Law of Gravitation implies that the two bodies of masses M and m, and the radius vectors $\vec{r}_M$ and $\vec{r}_m$, respectively, would exert the following forces of mutual gravitational attraction:

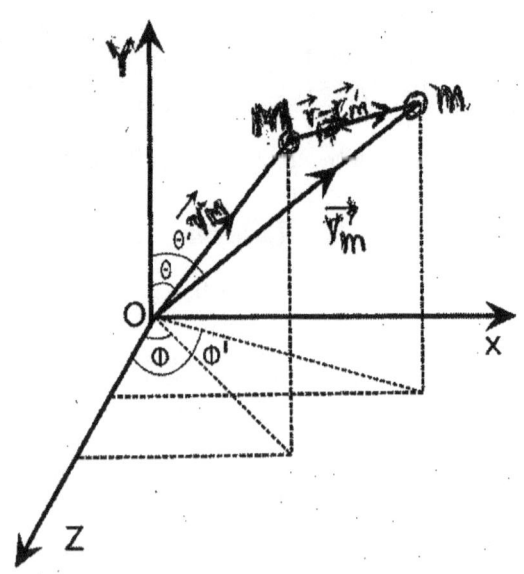

Figure 1.2. A two-body interaction coordinate system.
[Two masses M and m at the radius vectors r_M and r_m, respectively.]

Force on m: $\vec{F}_m = -m\ddot{\vec{r}}_m = -\dfrac{GMm}{r^2}\dfrac{\vec{r}}{r}$ (1.5a)

Force on M: $\vec{F}_M = -M\ddot{\vec{r}}_M = \dfrac{GMm}{r^2}\dfrac{\vec{r}}{r}$ (1.5b)

where G is the universal gravitational constant, and $\vec{r}$ is the vector joining M and m, taken +ve when directed from m to M (Figure 1.2). In the above equations, an arrow on top of a symbol denotes a vector quantity; one dot over it represents the First Order differential with respect to (wrt) time, t ; while two dots represent the Second Order differential with respect to time. The +ve and –ve signs on the right of Equations (1.5a,b) are consistent with the direction of $\vec{r}$ *to properly represent an attractive force* . Clearly, by *Triangle Law of vectors*:

$$\vec{r} = \vec{r}_m - \vec{r}_M$$ (1.5c)

Adding Equation (1.5b) and Equation (1.5a), we get:

$$\ddot{\vec{r}}_m - \ddot{\vec{r}}_M = \dfrac{G(M+m)\vec{r}}{r^2}\dfrac{\vec{r}}{r}$$ (1.5d)

For simplicity we assume M >> m, so that M is practically static in comparison to the motion of m under the gravitational force of M, we can simply assume $\ddot{\vec{r}}_M = 0$; so that, from Equation (1.5d),

$$\ddot{\vec{r}}_m = \dfrac{\beta}{r^2}\vec{r}_o$$

$$\vec{r}_o = \dfrac{\vec{r}}{r}$$ (1.5f)

where $\beta = G(M+m) \approx GM$ for $m \ll M$, and $\vec{r}_o$ is a unit vector along $\vec{r}$, Note that, for the gravitational force on the satellite (m) by the Earth (M):

$G = 6.672 \times 10^{-13}$ N cm²/g²

$M = 5.974 \times 10^{27}$ g

$GM = \beta = 3.986005 \times 10^{20}\ cm^3/sec^2$ \hfill (1.5g)

For further simplification, we assume that the origin of the coordinate system is chosen to at the (static) location of M, and the vector $\vec{r}_o$ makes an angle α wrt the x-axis. Then in terms of the Cartesian coordinates (x, y, z), $r = (x^2 + y^2)^{1/2}$; and with unit vectors $\hat{\imath}, \hat{\jmath}, \hat{k}$, along x -, y -, and z - axes, respectively,

$$\vec{r}_o = \cos\alpha\,\hat{\imath} + \sin\alpha\,\hat{\jmath} \hfill (1.6a)$$

The unit vector $\vec{r}_p$ perpendicular to $\vec{r}_o$, and in the direction of increasing α is:

$$\vec{r}_p = \hat{k} \times \vec{r}_o = -\sin\alpha\,\hat{\imath} + \cos\alpha\,\hat{\jmath} \hfill (1.6b)$$

Differentiating equation (1.2a) and (1.2b) with respect to time:

$$\dot{\vec{r}}_o = -\sin\alpha\,\dot{\alpha}\,\hat{\imath} + \cos\alpha\,\dot{\alpha}\,\hat{\jmath} = \dot{\alpha}\,\vec{r}_p \hfill (1.7a)$$

$$\vec{\dot{r}}_p = -\cos\alpha \cdot \dot\alpha \hat{i} - \sin\alpha \cdot \dot\alpha \hat{j} - \dot\alpha \vec{r}_o \qquad (1.7b)$$

Differentiating $\vec{r} = r\vec{r}_o$, the velocity of m could be written as:

$$\vec{v} = \vec{\dot{r}} = \dot{r}\vec{r}_o + r\vec{\dot{r}}_o$$

$$= \dot{r}\vec{r}_o + r\dot\alpha \vec{r}_p \qquad (1.7c)$$

Differentiating with respect to time again, we obtain the acceleration that m undergoes:

$$\vec{f} = \vec{\ddot{r}} = \vec{\dot{v}}$$

$$= \ddot{r}\vec{r}_o + \dot{r}\vec{\dot{r}}_o + \dot{r}\dot\alpha \vec{r}_p + r\ddot\alpha \vec{r}_p + r\dot\alpha \vec{\dot{r}}_p$$

i.e.,

$$\vec{f} = (\ddot{r} - r\dot\alpha^2)\vec{r}_o + (r\ddot\alpha + 2\dot{r}\dot\alpha)\vec{r}_p \qquad (1.8a)$$

We now confine to the motion of m along $\vec{r}_o$. Using Newton's Second Law of Motion, the equation of motion (along $\vec{r}_o$) can therefore be written by combining equation (1.5e) and equation (1.8a) as:

$$\vec{F_m} = m\vec{f} = m\vec{\ddot{r}}$$

Consequently,

$$\ddot{r} - r\dot{\alpha}^2 = -\frac{\beta}{r^2} \qquad (1.8b)$$

Also, since there is no motion (or force) in the perpendicular direction, the coefficient of $\vec{r}_p$ in Equation (1.8a) must vanish; i.e.,

$$\frac{1}{r}\frac{d}{dt}(r^2\dot{\alpha}) = r\ddot{\alpha} + 2\dot{r}\dot{\alpha} = 0 \qquad (1.8c)$$

Putting:

$$r = \frac{1}{u} \qquad (1.9a)$$

we have:

$$\dot{r} = -\frac{1}{u^2}\dot{u} = -\frac{1}{u^2}\frac{du}{d\alpha}\dot{\alpha}$$

$$= -r^2\dot{\alpha}\frac{du}{d\alpha} = -A\frac{du}{d\alpha} \qquad (1.9b)$$

where

$$A = r^2\dot{\alpha} \qquad (1.9c)$$

is the angular momentum per unit mass (i.e., $mA = mr^2\dot{\alpha}$ is the angular momentum of m), which is a constant.

Differentiating Equation (1.9b) with respect to time,

$$\ddot{r} = -A\frac{d}{dt}\left(\frac{du}{d\alpha}\right) = -A\frac{d}{d\alpha}\left(\frac{du}{d\alpha}\right)\frac{d\alpha}{dt}$$

$$= -A^2 u^2 \frac{d^2 u}{d\alpha^2} \qquad (1.9d)$$

where we have substituted for $\dot{\alpha}$ from Equation (1.9c). Substituting for $\ddot{r}$ from Equation (1.8b) and using equations (1.9a) and (1.9c) and simplifying, we obtain the equation of motion for m:

$$\frac{d^2 u}{d\alpha^2} + u = \frac{\beta}{A^2} \qquad (1.10a)$$

Solving the second order differential equation (1.10a), we can write:

$$u = B \cos(\alpha - \alpha_0) + \frac{\beta}{A^2} \qquad (1.11a)$$

where B and α_0 are the two integration constants with values depending on the initial conditions. Hence the equation of the orbit can be written as (Equation 5a)

$$r = \frac{p}{1 + e \cos(\alpha - \alpha_0)} \qquad (1.11b)$$

where

$$p = \frac{A^2}{\beta} \qquad (1.11c)$$

and

$$e = \frac{A^2}{\beta} B = p B \qquad (1.11d)$$

Equation (1.11d) is the polar equation of a conic section with p as the latis-rectum and e the eccentricity, with M as the focus, and the orbit of m being:

(i) a circle with radius p for $e = 0$
(ii) an ellipse   for $e < 1$
(iii) a parabola   for $e = 1$
(iv) a hyperbola   for $e > 1$

The cases of circular ($e = 0$) and elliptical ($e < 1$) orbits are of most interest for a satellite orbiting around the earth. The semi-major and semi-minor axes of the ellipse, a and b, respectively, are given as:

$$a = (d_1 + d_2)/2$$

$$b = a(1 - e^2)^{1/2}$$

where d_1 and d_2 are the distances of any point on the ellipse from the two foci, and their sum ($d_1 + d_2 = 2a$) is a constant for the ellipse. By a suitable orientation of the axes (x, y), we can choose $\alpha_o = 0$. Also, for an ellipse,

$$p = a(1 - e^2)$$

is the latis-rectum. Thus the equation of the orbit becomes:

$$r = \frac{a(1-e^2)}{1 + e \cos \alpha} \qquad (1.12a)$$

The minimum and maximum values of r are obtained by setting $\cos \alpha = 1$ and $\cos \alpha = -1$, (i.e., $\alpha = \pi$), respectively, and are given as:

$$min. r = r_p = a(1-e), \text{ (for } \alpha = 0) \qquad (1.13a)$$

$$max. r = r_a = a(1+e), \text{ (for } \alpha = \pi) \qquad (1.13b)$$

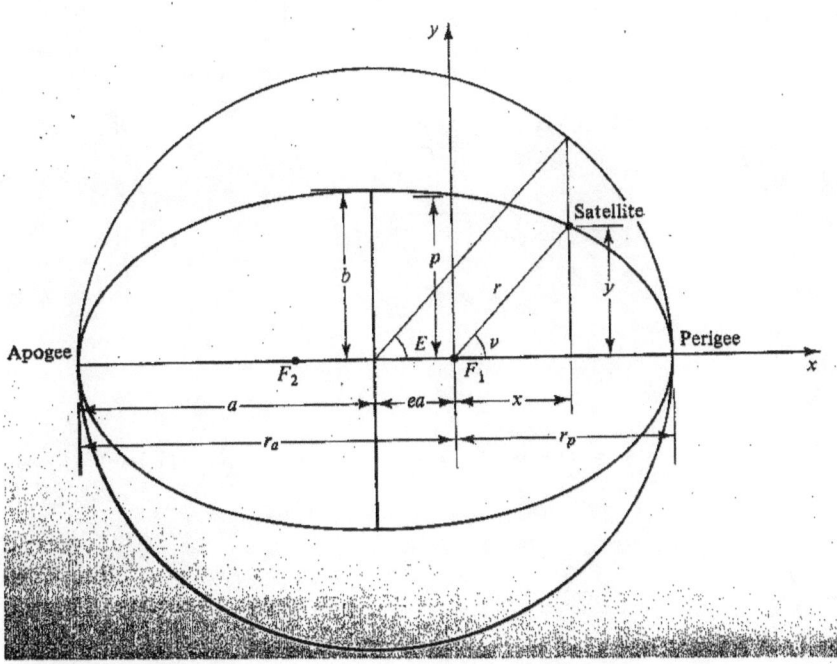

Figure 1.3. The orbital parameters under a central attractive force field.

[latis-rectum, $p = a(1-e^2)$; $F_1, F_2 \rightarrow$ Foci; a $\rightarrow$ semi-major axis; b $\rightarrow$ semi-minor axis.]

The above values are generally termed as *periapsis* and *apoapsis*, respectively. When Equation (1.12a) is with respect to the sun as the central body (i.e., for the case of planets revolving around the sun), the above minimum and maximum r-values are called the *perihelion* and *aphelion*, respectively; while when the central body is the earth (i.e., for terrestrial satellites), these values are referred to as *perigee* and *apogee*, respectively. Equation (1.12a), in its general form, represents an ellipse (E) with eccentricity, semi-major axis = $(r_a + r_p)/2 = a$; and semi-minor axis = $b = a\sqrt{1-e^2}$. The ellipse becomes a circle of radius a under $e = 0$; this circle is called the eccentric circle. The angular variable E (Figure 1.3) is called the *eccentric anomaly*, while the polar angle α with respect to the focus F_1 of the ellipse is called the *true anomaly*. If x and y are the Cartesian coordinates corresponding to the polar coordinates (r, α), the following relations can be easily obtained:

$$x = r \cos \alpha = a(\cos E - e) \quad (1.14a)$$

$$y = r \sin \alpha = a\sqrt{1-e^2} \sin \alpha \quad (1.14b)$$

so that

$$r = \sqrt{x^2 + y^2} = a(1 - e \cos E) \quad (1.14c)$$

Also,

$$\sin \alpha = \frac{(1-e2)^{1/2} \sin E}{1 - e \cos E} \quad (1.14d)$$

$$\cos \alpha = \frac{\cos E - e}{1 - e \cos E} \qquad (1.14e)$$

so that

$$\tan \alpha = \frac{\sqrt{1-e^2}\sin E}{\cos E - e}$$

Also,

$$\tan^2 \frac{\alpha}{2} = \frac{1-\cos\alpha}{1+\cos\alpha} = \frac{(1+e)(1-\cos E)}{(1-e)(1+\cos E)}$$

$$= \frac{(1+e)}{(1-e)} \tan^2 \frac{E}{2}$$

i.e.,

$$\tan \frac{\alpha}{2} = \sqrt{\frac{1+e}{1-e}} \tan \frac{E}{2} \qquad (1.14f)$$

The area R of the ellipse is:

$$R = \pi ab = \pi a^2 \sqrt{1-e^2} \qquad (1.15a)$$

Consequently, the constant *areal* velocity is given as:

$$\frac{dR}{dt} = \frac{R}{T} = \frac{\pi a^2 \sqrt{1-e^2}}{T} \qquad (1.15b)$$

where T is the period or time taken to complete one revolution. Also, from conservation of angular momentum,

$$\frac{dR}{dt} = \frac{1}{2}\sqrt{\beta p} = \frac{1}{2}\sqrt{\beta a(1-e^2)} \qquad (1.15c)$$

Hence equating the rhs of Equations (1.11b) and (1.11c), we obtain the result confirming Kepler's Third Law:

$$T^2 = \frac{4\pi^2}{\beta} a^3 \qquad (1.15d)$$

or

$$T = 2\pi \sqrt{\frac{a^3}{\beta}} \qquad (1.15e)$$

and the parameter

$$n = \frac{2\pi}{T} = \sqrt{\frac{\beta}{a^3}} \qquad (15f)$$

is referred to as the *mean motion*. The equation of the ellipse in the Cartesian coordinates with origin at the focus F_1 is clearly:

$$\frac{(x-ea)^2}{a^2} + \frac{y^2}{b^2} = 1 \qquad (16a)$$

1.6. Basic Types of Satellite Orbits and Applications

1.6.1. Geostationary Satellites (GEO)

The GEO satellites occupy circular orbits at a fixed altitude of R_G above the earth. They are synchronous with respect to the daily rotation of the earth about its axis, and hence appear stationary for an earth-based antenna pointed to such a GEO (also called *geosynchronous*) satellite. In contrast, a satellite with altitude smaller than R_G is called a *low earth orbit* (LEO) satellite. Medium altitude satellites are called *medium earth orbit* (MEO), and similarly for *highly elliptical earth orbit* (HEO) satellites. Communications satellites are GEO satellites which, according to Kepler's Third Law (Equation 1.15d), satisfy the relation

$$R_o = \left(\frac{\beta}{4\pi^2} T^2\right)^{1/3} \tag{1.17a}$$

where R_o is the radius of the orbit from the center of the earth. Substituting (Equation 1.5g)

$$\beta \cong GM = 398600.5 \; km^3/s^2$$

and

$$T = 86164.1 \; s$$

we obtain

$$R_o = 42164.2 \; km \qquad (1.17b)$$

Subtracting the radius of the earth, the altitude of a GEO satellite above the equator is obtained as:

$$R_G = 35{,}786 \; km \qquad (1.17c)$$

which is approximately 6 times the earth's radius. The velocity of the GEO satellite then becomes

$$v = \frac{2\pi R_G}{T} \simeq 3.075 \; km/s \qquad (1.17d)$$

1.6.2. Non-Geostationary Satellites

For many applications, especially those involving relatively weak signals (e.g., mobile communications with small terminals), it is advantageous to have satellites with low-altitude (i.e., low earth orbit – LEO satellites), since then the propagation loss is smaller compared to geostationary (GEO) satellites. It is then natural that the (LEO) satellites have a faster angular velocity than the period of the diurnal rotation of the earth; so the satellite changes its position relative to the earth's surface due to revolution around the earth. It is usually of interest to trace the path of the satellite upon the earth surface, which can be done for given orbital parameters of the satellite. For certain applications, a circular orbit around the earth may be instituted; while for others, an elliptical orbit may be preferred.

For the sake of generality, it suffices to consider a satellite with an elliptical orbit with a specified eccentricity (e) and period (T). The case of a circular orbit is readily derived as a special case by setting the eccentricity to be zero ($e = 0$).

Kepler's familiar laws for orbital motion in an elliptical orbit are used to derive various properties of the satellite orbital motion, including a ground trace.

For the pertinent analysis, the position of the satellite on a non-rotating earth is obtained (i.e., in the inertial geocentric coordinates), and then the earth's rotational motion is introduced to trace the satellite motion on the ground for a realistic case.

The satellite's period (in an inertial frame corresponding to the earth – based coordinates which are fixed in this frame, assuming non-rotating earth) is related to the mean motion and the semi-major axis (Kepler's Third Law) as:

$$T = 2\pi\sqrt{\frac{a^3}{\beta}} = \frac{2\pi}{n} \qquad (1.18a)$$

where

$T =$ satellite's orbital period
$a =$ semi-major axis of the elliptical orbit
$n =$ mean motion (angular velocity)
and $\beta = GM = 398{,}600.5 \ km^3/sec^2$

G being the universal gravitational constant and M, the mass of the Earth.

The mean anomaly (M) at a particular time t, the eccentric anomaly (E), and the true anomaly (θ) can then be written as:

$$M = n(t - t_o) + M_o \qquad (1.19a)$$

$$E = M + e \sin E \qquad (1.19b)$$

and

$$\theta = 2\tan^{-1}\left[\left(\frac{1+e}{1-e}\right)^{1/2}\tan\left(\frac{E}{2}\right)\right] \quad (1.19c)$$

where M_o is the mean anomaly at a given initial time (epoch) t_o; Equation (1.19b) results from Kepler's Equation, while Equation (1.19c) from the so-called Gauss' Equation for orbital motion.

The satellite's position in its orbit, (r, θ), can then be computed as follows.

We observe that (see Figure 1.3):

$$x = a\cos E - ea = a(\cos E - e)$$

$$= r\cos\theta \quad (1.20a)$$

$$y = a\sqrt{1-e^2}\sin E$$

$$= r\sin\theta \quad (1.20b)$$

Hence,

$$r = \sqrt{x^2 + y^2} = a[(\cos E - e)^2 + (1-e^2)\sin^2 E]^{1/2}$$

$$= a[1 + e^2(1-\sin^2 E) - 2e\cos E]^{1/2}$$

i.e.,

$$r = a(1 - e\cos E) \qquad (1.20c)$$

$$\tan\theta = \frac{y}{r} = \frac{\sqrt{1-e^2}\sin E}{1-e\cos E} \qquad (1.20d)$$

Alternatively, the orbital polar angle θ can be written as (see Equation 1.20a):

$$\tan^2\frac{\theta}{2} = \frac{1-\cos\theta}{1+\cos\theta} = \frac{1-\frac{a}{r}(\cos E - e)}{1+\frac{a}{r}(\cos E - e)}$$

i.e.,

$$\tan^2\frac{\theta}{2} = \frac{r - a(\cos E - e)}{r + a(\cos E - e)} \qquad (1.21a)$$

Substituting for r from Equation (1.20c), we get:

$$\tan^2\frac{\theta}{2} = \frac{(1-e\cos E) - (\cos E - e)}{(1-e\cos E) + (\cos E - e)}$$

$$= \frac{(1+e) - \cos E(1+e)}{(1-e) + \cos E(1-e)}$$

$$= \frac{(1+e)(1-\cos E)}{(1-e)(1+\cos E)}$$

$$= \left(\frac{1+e}{1-e}\right)\frac{\sin^2 E/2}{\cos^2 E/2}$$

i.e.,

$$\tan\frac{\theta}{2} = \left(\frac{1+e}{1-e}\right)^{1/2} \tan\frac{E}{2} \qquad (1.21b)$$

For a circular orbit ($e = 0$), we have the polar coordinates (r_o, θ_o) given from equations (1.20c) and (1.21b) as expected.

$$r_o = a \qquad (1.21c)$$

$$\theta_o = E \qquad (1.21d)$$

Introducing now the rotation of the earth along θz about its axis, the Cartesian coordinates (x', y', z') corresponding to the (non-rotating) coordinates (x, y, z) can be written in terms of the (2 x 2) earth's rotation matrix (assuming spherical earth), R',

$$\begin{bmatrix} x' \\ y' \end{bmatrix} = R' \begin{bmatrix} x \\ y \end{bmatrix} \qquad (1.22a)$$

where, R', the 2 x 2 Rotation matrix is given as

$$R' = \begin{bmatrix} \cos w(t-t_o) & \sin w(t-t_o) \\ -\sin w(t-t_o) & \cos w(t-t_o) \end{bmatrix} \qquad (1.22b)$$

w being the earth rotational (angular) speed, and the rotation is assumed to be with the z-axis as the axis of rotation; and t is the current time, with t_o as the initial time (epoch), as mentioned before.

i.e.,

$$R' = \begin{bmatrix} \cos w\tau & \sin w\tau \\ -\sin w\tau & \cos w\tau \end{bmatrix}$$

where (1.22c)

$$\tau = t - t_o$$

is the time elapse from the epoch. Thus,

$$x' = x \cos w\tau + y \sin w\tau$$

$$y' = -x \sin w\tau + y \cos w\tau \qquad (1.22d)$$

Substituting for x and y from equations (1.20a) and (1.20b), we have

$$x' = r \cos\theta \cos w\tau + r \sin\theta \sin w\tau$$
$$= r \cos(\theta - w\tau)$$
$$y' = -r \cos\theta \sin w\tau + r \sin\theta \cos w\tau$$
$$= r \sin(\theta - w\tau) \qquad (1.22e)$$

Consequently,

$$r' = (x'^2 + y'^2)^{1/2} = (x^2 + y^2)^{1/2}$$

$$= r$$

and

$$\tan\theta' = \frac{y'}{x'} = \tan(\theta - w\tau)$$

The polar coordinates of the point $P'(r', \theta')$ on the earth corresponding to the point $P(r,\theta)$ in the orbital plane can then be expressed in terms of the latitude θ_E and longitude λ_E as

$$\theta_E(t) = \theta$$

$$\lambda - \lambda_o(t_o) + w\tau$$

where $\lambda_o(t_o)$ is the initial epochal longitude.

Determining the θ_E and λ_E at successive instants of time, the ground-trace of the satellite can be obtained.

1.6.3. Sun-Interference

In course of the annual revolution of the earth in its orbit around the sun, and the diurnal rotation of the earth around its axis, the motions also shared by a geostationary satellite and other types (LEO) satellites, there are occasions when the sun appears to come in the line of sight for the antenna of an earth-station which is pointed to the satellite to receive its signal. This situation may last for a short period of time, during which the solar radiation directly impinges upon the receiving earth station antenna, causing a large amount of noise, essentially obliterating or overwhelming the signal altogether; thereby causing an outage. This type of outage of the earth station is more pronounced for large antennas, which correspond to a narrow beam. The sun subtends an angle of about 0.5° at a point on the earth, so for an antenna with a beam width of about 0.5° or less, the apparent antenna noise temperature can become comparable to the characteristic solar temperature during this conjunction or sun-interference.

As the sun is closest to the equator at equinox, for a geostationary satellite and equatorial earth station, the sun-interference is most severe, and its duration is longest when sun-interference takes place at equinox.

The formulae to calculate the time of occurrence of sun-interference are provided below. First the relevant variables and the notations used here are defined:

r_e = the radius of the earth
r_s = the radius of the satellite (= 42,164 km for geostationary or GEO satellite)
λ_s = longitude of the (GEO) satellite
λ_e = longitude of the earth station
θ_e = latitude of the earth station
R = the range (distance of the earth station and the satellite)
$\Delta \lambda$ = difference between the two longitudes

$$= \lambda_s - \lambda_e$$

δ = declination of the sun
ϵ = inclination of the earth's orbit with respect to the ecliptic
α = sun's right ascension

β = angle between the two straight lines, one joining the earth's center and the earth station, and the other joining the earth station and the satellite

E = elevation of the satellite

A = azimuth of the satellite

H = hour angle

⋮

Using the properties of spherical triangles, we can write the following relations:

$$\beta = \cos^{-1}(\cos\theta_e \cos\Delta\lambda) \quad (1.23a)$$

$$R = [r_e^2 + r_s^2 - 2r_e r_s \cos\beta]^{1/2} \quad (1.23b)$$

$$A = \sin^{-1}\left(\frac{\sin\Delta\lambda}{\sin\beta}\right) \quad (1.23c)$$

$$E = \cos^{-1}\left(\frac{r_s}{R}\right)\sin\beta \quad (1.23d)$$

$$A = \sin^{-1}(\sin D1/\sin\beta) \quad (1.23e)$$

$$\cos H = \frac{\sin E - \sin\theta_e \sin\delta}{\cos\theta_e \cos\delta} \quad (1.23f)$$

$$\tan\alpha = \frac{\cos E}{\sqrt{\sin^2 E/\sin^2\delta - 1}} \quad (1.23g)$$

Finally, the Local Mean Sidereal Time L, for the sun interference is given as:

$$L = H + \alpha \quad (1.23h)$$

References:

[1] W. L. Pritchard, H. G. Suyderhoud, and R. A. Nelson, Satellite Communication Systems Engineering, Prentice Hall, Englewood, N.J. 1993.

[2] B. N. Agrawal, Design of Geosynchronous Spacecraft, Prentice Hall, Englewood Cliffs, NJ, 1986.

2. MAJOR SYSTEM COMPONENTS (SUBSYSTEMS)

In this book, emphasis is laid only on the *theory* of satellite communications (SATCOM), including mobile telecommunications. It therefore focuses on presentation of theoretical formulations of the various processes, and mathematical representation and analyses of some of the basic functionalities involved. A fairly vast and important area of SATCOM not dealt with in this book for the sake of brevity includes description of various physical systems and hardware equipment employed in the satellite and the earth stations – their engineering designs, their operational and implementation details, their construction and optimization processes, relative cost considerations, and so on.

It is desirable, however, at least just to mention some of the important equipment in the satellite (space segment) and the earth stations (earth segment), together with a few words regarding their relevance and applications in a brief qualitative manner, for a theoretical analyst not familiar with such a background. This Chapter is an attempt in that direction. Details in this connection can be found in any good book on SATCOM.

2.1. Satellite (Space Segment)

The assembly of equipment, properly designed and integrated, that is launched in the pre-designated orbit with the help of a suitable rocket, constitutes the satellite commonly and collectively referred to as the space segment. Prior to ejection from the upper stage of the rocket, the satellite (or a set of satellites) is referred to as the *payload* of the rocket. The selection of the type of the rocket system is determined on the basis of the overall weight of the payload and the altitude at which the satellite(s) need to be orbited (GEO, MEO, LEO, etc.) Conversely, for each class of rocket, there are limitations and constraints imposed on the weight of the payload and the altitude that can be achieved by them. The basic components of the satellite usually include the repeater assembly, solar panels, battery system, structure and attitude control system, and, last but not the least, antenna(s).

2.1.1. Repeaters or Transponders

Repeaters or transponders provide amplification of the power level of the input signal to yield an output signal tens of dBW (watts, W, expressed in *decibles*, i.e., $10\log_{10}W$) higher in power level. This amplification is achieved by means of energy exchange from an energetic electron beam to the electromagnetic wave or carrier.

The transponder can perform the above function in a transparent manner, without any decomposition and re-composition of the input carrier onboard. In this case the main function of the transponder remains appropriate level of power amplification and then a frequency translation from the uplink frequency band to a downlink frequency band. Alternatively, the transponder can first demodulate the received signal to restructure it based on the grouping of transmitting and receiving earth stations, in addition to providing the requisite power amplification. The latter process is usually employed in the case of digital signals, for which the baseband signal can be accurately regenerated and coded to obtain a robust and secure link. In the case of mobile telecommunications by satellite and terrestrial networks, the modulation and multiple access of the digital signal is commonly based on single-channel-per-carrier and code division multiple access (SCPC/CDMA) digital transmission with regeneration of the signal (baseband) bits (The implications of some of the terms used here will become clear in the later Chapters). Digital television transmission also benefits from bit regeneration to achieve higher picture quality. In general, digital voice including mobile telephony uses SCPC/CDMA or multiplexed voice channels depending on system requirements. Satellite radio also is broadcast digitally.

The main varieties of repeaters are as follows.

2.1.1.1. Traveling Wave Tube Amplifier (TWTA)

As the name implies, TWTAs achieve amplification of the input wave by means of exchange of energy while the wave is traveling along with a beam of energetic electrons along a tubular structure. The electrons are produced from a thermionic source (e.g., cesium) heated appropriately to yield a copious supply of electrons, which are allowed to traverse the length of the tube and are finally collected at the other end by means of collectors which are kept at a "depressed" potential to optimize the collection. The depressed collector may be constructed and operated over a multi-stage manner for such optimization, so as to facilitate collection of all the "spent" electrons which may have a range of residual energy. Suitably placed magnets along the tube help the beam to remain focused instead of being dispersed and scattered by the tube-walls.

As the wave travels along a helical conducting wire so as to match with the electron beam velocity, it has adequate time to interact with the electron beam and extract its energy and become amplified in power. The slow-wave structure may be implemented for optimizing the energy-exchange and to attain the required level of power amplification for the carrier wave, which is finally output at the other end of the TWTA. The wave then passes through a mixer for frequency translation into the downlink frequency band, and transmitted down to receive earth stations via the downlink satellite antenna. Figure 2.1 shows the basic schematics of a TWTA.

```
                              I
                TWTA Wall     V
_____
I  I----------------------------------->-----------------IDepressed
I  I-------------Electron Beam-------------->-------------------------I
I  I------------------------------------------------>-----------ICollector
^           Slow Traveling Wave --->
I ___I input wave I_____I output wave I____
I       ^            TWTA Wall      I      V      I
Thermionic  ^                              V
Source (of electrons)
```

Figure 2.1. The Basic Structure and Schematics of the TWTA

2.1.1.2. Solid Sate Power Amplifier (SSPA)

As an alternative to the conventional TWTA, solid state power amplifiers (SSPAs) could be used in the satellite to achieve power amplification of the input wave thousand-, even million-fold. In this case, instead of relatively bulky TWTAs, energetic electron generation and slow wave structure comprises solid state circuitry. Due to the inherent advantages of compact and light-weight solid state microelectronic components, SSPA is a preferred choice, especially when satellite (payload) weight is a critical issue.

In any case, depending on traffic and capacity requirements, a large number of TWTAs or SSPAs are used in a satellite. Each power amplifier and mixer assembly (i.e., transponder) is properly connected to the appropriate antennas to receive the input uplink carrier and to transmit the output downlink carrier.

2.1.2. Solar Panels

Solar panels, attached to the body of the satellite, are folded for compactness in the payload, and these panels are unfurled after the satellite achieves proper orbital position and orientation (Attitude). The overall dimensions and weight of the solar panels are of course planned in consistence with the constraints of the selected rocket system on the one hand, and in order to meet the power generation requirement for the satellite mission over its lifetime, on the other.

The solar panels usually comprise a set of hinged surfaces covered with thousands of solar cells that generate electric power from the solar radiation incident upon them. The solar panels are therefore oriented properly to optimize the amount of power generated. The cells are connected in series to accumulate the electric power. The electric power thus generated activates various other components of the satellite as required.

The solar cells are selected on the basis of their intrinsic efficiency for power generation out of the solar radiation, as well as of their weight, among other properties (reliability, lifetime, cost, etc.) The assembly of the solar cells is protected from other radiations in space (particles in the magnetosphere, cosmic rays) by means of light-weight coatings. Typically, the solar cells are of photovoltaic type, for example, Gallium-Arsenic (GaAs) crystals. Improvements in the power-generation efficiency and weight-reduction, cost reduction, etc. to optimize the use of the solar cells are ongoing process. The solar cell power generation technology as a spin-off industry from satellite and space technology has now become useful for terrestrially-based solar power generation for various applications including electricity for residential and office buildings. On the other hand, solar power satellites for large-scale generation of power in space for transmission to the earth through a laser beam has also been considered and may become a reality with further technological advances.

2.1.3. Battery System

A suitable power storage system, i.e. a battery, is required in the spacecraft when the direct solar radiation for generating electric power becomes unavailable to the satellite, that is, during a solar eclipse, in course of the orbital motion of the satellite. The battery is charged using the solar radiation power during non-eclipse period, and it becomes gradually discharge during the period of solar eclipse as it provides the requisite level of power. The battery is recharged in the next phase when the solar radiation becomes available once again (i.e., when the satellite comes out of the solar eclipse condition). The battery provides the power in the next cycle of eclipse period. Thus this charging-and-discharging process continues periodically throughout the operational life of the satellite.

The strength or quality of the battery, apart from efficiency, weight, and cost, is commonly designated by the parameter called "discharge-depth" which determines how many times the recharging process can be carried out effectively (and hence the life of the battery, impacting the mission of the satellite). The discharge-depth of the battery employed is one of the important parameters as a part of the overall specifications of the satellite electrical subsystem.

2.1.4. Attitude Control Subsystem (ACS)

After the satellite is launched to occupy the desired orbit, it has to be oriented properly so that its uplink and downlink antennas accurately cover their pre-designated coverage areas; and its solar panels are also oriented to optimally receive the solar radiation. Other orientation-based requirements may include a heat radiation into the space as part of the cooling device. Such orientation determination and control is provided by its attitude control subsystem (ACS) built in the satellite design.

The ACS usually comprises a spinning flywheel which controls and preserves the satellite attitude by virtue of the conservation of angular momentum. Alternatively, a system of the so-called Three-Axis attitude control is used. Sensors for the sun, star, and earth are employed to help achieve the proper orientation or attitude for the satellite.

2.1.5. Satellite Antennas

The satellite is usually equipped with two or more antennas which are also stowed in the spacecraft during the launch phase, and are deployed when the satellite achieves the desired orbit and attitude. Typically, one or more antenna(s) are used to receive the uplink carrier from the transmitting earth station; while one or more separate antennas are used to provide transmission of the downlink down to the receive earth stations. All the antenna sizes and types are selected to provide the necessary geographical coverage of the intended regions of the earth, as well as the required antenna gains to attain the desired signal quality.

Different uplink and downlink frequency bands are of course required to avoid interference between the respective carriers. A set of different antennas with appropriate sizes and efficiencies may be required depending on the multiple geographical coverage regions, or on the use of orthogonally polarized waves to achieve frequency reuse for higher capacity. Accurate antenna pointing toward its coverage region is a part of the preparatory maneuvers for the satellite before operation can start.

Satellite telecommunication (SATCOM) was initiated by INTELSAT (International Telecommunications Satellite Organization) by using global-coverage (covering one-third of the earth) antennas operating in the C-Band frequencies: 6 GHz for uplink and 4 GHz for the downlink (denoted by 6/4 GHz frequency band). This frequency band is least affected by atmospheric transmission and noise due to the cosmic radiation. As this 6/4 GHz band became saturated, the 14/11 GHz Ku-Frequency Band, and subsequently, the 30/20 GHz Ka-Frequency Band were employed, to successively achieve higher and higher satellite capacity. These higher frequency bands are utilized to obtain hemispheric (roughly 1/6the of the earth surface) and smaller spot beams for coverage regions. The antenna size becomes progressively smaller as higher frequency bands are employed; although problems due to atmospheric and rain absorption increase.

In mobile telecommunications, a multiplicity of small spot coverage regions is involved.

Often, hexagonal-shaped contiguous spot coverage regions forming an overall geographical region are generated for this purpose. Consideration of interference among neighboring hexagonal spots must be considered for system design in such cases.

2.1.6. Thermal Control Subsystem (TCS)

The TCS is designed to prevent excessive heating or excessive cooling of the satellite – conditions which would likely affect the operation of other subsystems. For example, for optimal performance, the battery subsystem must be kept within a narrow temperature range. Usually a radiation cooling device releasing excess heat generated within the satellite into free space is employed.

2.2. Earth Stations (Earth Segment)

Satellite communications (SATCOM) essentially entails transmission(s) from one or more transmitting earth stations to the satellite (uplink) and transmission from the satellite to one or more receiving earth stations, including broadcast to a large number of receiving earth stations (downlink). The earth stations may be situated within the global, hemispheric, or spot beams of the satellite antenna(s) covering the corresponding geographical region(s). The individual earth stations are equipped with antenna(s) for uplink transmission to, or downlink reception from, the satellite. In addition, the major earth station equipment include high power amplifier (HPA) or low noise amplifier (LNA) to perform their transmitting and receiving functions, respectively.

2.2.1. Earth Station Antenna

Usually the transmitting earth station has a relatively very large antenna necessitated by its high transmission rate and capacity required. The antenna is mounted on a pedestal and can be oriented for tracking the satellite employing sophisticated tracking mechanism. The larger value of the transmit antenna diameter implies a correspondingly narrow transmit or uplink beam, so that tracking and pointing accuracy becomes crucial.

The receiving earth station antenna size could be large or small as required by the receive traffic capacity and signal quality requirement. Depending on the antenna size, pointing to the satellite or tracking can be automatic or manual. Comparatively, a broader antenna beam is associated corresponding to a smaller antenna diameter, easing the pointing and tracking function.

The performance of any antenna is indicated by the ratio of the antenna gain (G) to the associated noise temperature (T) – i.e., by its **G/T**-value. Often, in a specific SATCOM network, one or more Standard antenna sizes and performance are specified to ensure signal quality and network integrity. For example, in the INTELSAT global network, the use of Standard A, Standard B, Standard C, and Standard D, among others, are utilized, each with a specified antenna size, G/T-value as well as other requirements (cross-polarization ratio or XPR, surface smoothness, off-axis radiation limits, etc.) These Standards were introduced in the network at different phases of development of the network and to meet different types of service applications introduced. In the case of the ECHOSTAR satellite network for digital television transmission and broadcast service*, the receive earth terminals are of small (about 45 cm, called Dish antenna) diameter to receive digital television transmission (broadcast) by the satellite. These "Dish-Network" antennas could be mounted on roof-tops of houses and buildings. The antenna cost of course depends on its size, and the Dish-antennas are rather inexpensive, making receiving the broadcast digital television affordable to a large population. In contrast, INTELSAT Standard A earth stations are quite expensive, costing about one million U.S. dollar or more. INTELSAT Standard D earth stations were introduced as involving a low-cost antenna for small amount of traffic of reasonable quality, called the VISTA Service*

The author (A. K. Sinha) was actively involved in the development of (the theoretical aspects of) the INTELSAT Standard D earth station, related antenna performance analysis and introduction of the VISTA Service, as the Service Development Manager of INTELSAT. Similarly, the author was responsible for the development of the electromagnetic performance specifications and analysis for the ECHOSTAR-I ("Dish Network") as a Consultant. It is gratifying to see the small Island countries of the South Pacific use Pacific-Region INTELSAT satellites for their domestic and international telecommunications needs; and also the Dish antennas atop roof-tops everywhere in the world as a part of the Dish-Network services (Digital TV broadcast). The VISTA service is of course also used in many remote and isolated region telecommunications requirements, such as the oil-rigs, rural telecommunications, etc.

– so that even small countries such as the small Island countries in the South Pacific region could afford it and avail domestic and international satellite telecommunications.

2.2.2. High Power Amplifier (HPA)

HPA is used in the transmitting earth station for adequate amplification of the uplink carrier power before its transmission to the satellite. The HPA is also of a slow-wave structure type to maximize power exchange from a thermionic source for producing energetic electron beam to the input carrier wave, so that the output carrier wave is more powerful by tens of decibels (dBW). As the earth station is not bound by any weight constraints, the slow wave structure typically used in a transmit earth station is of a type called the coupled-cavity travelling wave tube (CC-TWT), which is composed of a series of coupled cavities through which the input wave is made to travel to optimally interact with the electron beam. The resulting equipment (HPA) is relatively heavy, but is capable of higher level of power amplification than the conventional (helical wire) type.

2.2.3. Low Noise Amplifier (LNA)

As the name implies, this type of power amplifier is used in a receiving earth station to provide the highest signal-to-noise ratio **(S/N)**, or carrier-to-noise ratio **(C/N)**, or other related performance criteria (e.g., carrier-power-per-bit-to-noise-density ratio, E_b/N_o, in the case of digital communication). The noise is invariably produced due to random motion of electrons in the conductor involved in the receiver electronics, and is difficult to remove altogether; at best this thermal noise could be minimized for a better performance of the receiving earth station and, hence, for improved signal quality.

In addition to the above-mentioned major equipment, many other auxiliary equipment may be utilized in the satellite and in transmit and receive earth stations of the SATCOM network. These include, for example, multiplexers and de-multiplexers (MUX/DEMUX); coder and decoder (CODEC); diagnostic equipment; command and control and telemetry (TT&C) equipment. The reader is referred to a general text on SATCOM to see details of the equipment in a satellite telecommunications network, and the roles they play in determining the size and weight of the spacecraft payload, earth station network design, cost, and signal transmission performance; since, as mentioned before, this book concentrates only on the theoretical background of the system and its component functions.

A theoretical outline of how to compute the signal quality is provided in the next Chapter.

3. THE LINK EQUATION

3.1. General

The satellite usually interlinks two earth-stations for transmission of the signal from one to the other, or for broadcast to a set of earth stations within its coverage zone. The signal to be transmitted from an earth-station E1 to the satellite is referred to as the *uplink*. The satellite amplifies this uplink signal, translates its frequency band, and then transmits it to another station E2 (or broadcasts it to a group of earth-stations or over a geographical region). The signal re-transmitted from the satellite to the ground is referred to as the *downlink*. The signal is carried in space by an electromagnetic (EM) carrier wave of suitably chosen frequency, with an associated bandwidth. The carrier EM wave is appropriately modulated by the signal at the transmit end, and demodulated to recover the signal at the receive end. In general the modulation and demodulation processes can be performed using one of a number of techniques including analog or digital methods. In this book we limit our consideration only to digital methods for modulation/demodulation (modem) and coding/decoding (codec) processes and equipment. Modern systems and service applications commonly employ digital systems only on account of a multitude of advantages inherent in the digital signals and systems.

Apart from the desired signal, there inevitably exists some level of noise due to random thermal motion of the electrons in various equipment, and due to some extraneous sources of noise including background cosmic radiation, adjacent satellite radiation, adjacent channel signals, nonlinearity of the response function of the high power amplifier (HPA) at the transmit side, and of the low noise amplifier (LNA) at the receive side, and so on.

Various measures of the quality of signal transmission and of the performance include the following:

$$\frac{S}{N} = \text{signal-to-noise ratio} \qquad (3.1\text{a})$$

$$\frac{C}{N} = \text{carrier-to-noise ratio} \qquad (3.1\text{b})$$

$$\frac{E_b}{N_o} = \text{energy-per-bit-to-noise-power-density (for digital signals)} \qquad (3.1\text{c})$$

The evaluation of any measure of the quality of signal transmission, i.e., the level of system performance is performed systematically by using the various relevant system and sub-system parameters. This process is commonly referred to as the *link-budget calculation*.

Typically, the signal quality for uplink (say, from the earth station E1 to the satellite S) and downlink (from the satellite S to the earth station E2) each is evaluated separately, and the two results are then combined to obtain the net quality for the overall link from the earth-station E1 to the earth-station E2 via the satellite S. The main parameters entering the link-budget calculation include the transmit and receive antenna sizes of E1, E2 and S; the frequencies of the uplink and downlink carrier-waves and their bandwidths; uplink and downlink transmit powers; system (thermal) and interference noise power density (noise power per unit bandwidth); and the range(s) or distance(s) – between E1-to-S and S-to-E2. The derived quantities include the antenna gain; the so-called *equivalent isotropic radiated power (e.i.r.p.)*; the noise power; *path-losses*; and finally the C/N and E_b/N_o values.

The formulation of the link-budget calculation is outlined symbolically in the next section, together with an example with specified numerical values of the parameters. A software implementation scheme available in a publicly open website domain is then indicated by means of a flow-chart type of representation.

For convenience in numerical computation, it is customary to represent various quantities involved in *decibels* (dB), defined as follows:

$$x(dB) = 10 \log_{10}(x) \qquad (3.2a)$$

where x on the right of the equation is the absolute numerical value of the parameter in question. This conversion (into dB) allows multiplication and division to be carried out in the forms of addition and subtraction, respectively, by virtue of the well-known mathematical properties of the logarithm function; *viz.*, (omitting the base '10' for simplicity):

$$\log(x_1 x_2) = \log x_1 + \log x_2$$

$$\log\left(\frac{x_1}{x_2}\right) = \log x_1 - \log x_2$$

3.2. Link-Budget Formulation

3.2.1. Uplink

For the uplink, let the following symbols and notations be used:

f = frequency of the carrier EM wave (in GHz, ie, 10^9 cycles/second)
P = trasmit power (watts)
D = transmit antenna diameter (meters)

T = noise-temperature of transmit system (Kelvin)
η = efficiency of the antenna (%)
R = path-length, ie the range (distance) from the transmit earth-station (E1) to the satellite (S), (km)
B = bandwidth (Hz)
k = Boltzmann Constant = 1.3807×10^{-23} J/K (Joules/Kelvin)

Then, we have:

G = (transmit) antenna *Gain* (with respect to isotropic radiation)

$$= \left(\frac{\pi D}{\lambda}\right)^2 \eta \qquad (3.3a)$$

where

$$\lambda = \frac{c}{f \times 10^9} = \frac{2.998 \times 10^{10}}{f \times 10^9} = \frac{29.98}{f} \, cm \qquad (3.3b)$$

is the wavelength of the transmit EM wave, and $c = 2.998 \times 10^{10}$ cm/s is the velocity of light (i.e., of the EM wave).

The half-power beam width, θ, is given as:

$$\theta = \frac{\lambda}{D \times 10^2} \, (degrees) \qquad (3.3c)$$

It can be noted that:

R = 35,788 km at the sub-satellite point (3.3d)
or, R = 41,679 km at edge of earth coverage (3.3e)

In *dB*, the transmit antenna gain, with respect to isotropic radiation) can be written as:

$$G(dBi) = 10 \log\left[\left(\frac{\pi D}{\lambda}\right)^2 \eta\right] = 10 \log\left[\left(\frac{\pi D \times 10^2 \times f \times 10^{+9}}{2.998 \times 10^{10}}\right)^2 \eta\right]$$

$$= 20 \log(10.479 \times fD) + 10 \log \eta$$

$$= 20.4 + 20 \log f + 20 \log D + 10 \log \eta \qquad (3.4a)$$

where it must be recalled that f is in GHz and D in meters, and log is with base 10.

If the transmit power is radiated in an isotropic fashion, it would be equally distributed over the inner surface area $4\pi R^2$ of a sphere of radius R, resulting in a power density value $= \frac{P}{4\pi R^2}$, at the altitude of the satellite. But the transmit antenna enhances this value by a factor G.

Consequently, the effective transmit power density becomes

$$= \frac{P}{4\pi R^2} G$$

Similarly, the satellite uplink receive antenna gain, G_r, enhances the receive power by a factor G_r, giving a received power value

$$C = \frac{P}{4\pi R^2} G\, G_r \qquad (3.4b)$$

The system noise power can be written as:

$$N = (kT)B = N_o B \qquad (3.4c)$$

where $N_o = kT$ is the noise power density (i.e., noise power per unit bandwidth, k being the Boltzmann constant ($k = 1.3807 \times 10^{-16}$ ergs/K). Thus the carrier power-to-noise power becomes:

$$\frac{C}{N} = \frac{C}{kTB} = \frac{C}{N_o B} \qquad (3.5a)$$

Note that, in dB, $-10 \log k = 228.6\ dB$. Therefore, receive power-to-noise temperature ratio, $\frac{C}{T}$, is :

$$\frac{C}{T} = \frac{P}{4\pi R^2} G \frac{G_r}{T} = \left(\frac{PG}{4\pi R^2}\right) x \left(\frac{G_r}{T}\right) \qquad (3.5b)$$

or, in dBW/K, we have

$$\frac{C}{T}(dBW/K) = 10\log(PG) - 10\log(4\pi R^2) + 10\log\frac{G_r}{T} \qquad (3.5c)$$

Now the *equivalent isotropic radiated power (e.i.r.p.* or simply *eirp)* is defined as

$$\text{eirp} = 10\log(PG) = 10\log P + 10\log G \quad (dBW)$$

$$= 10\log P + G \quad (dBW) \qquad (3.5d)$$

assuming G is already expressed in dB; and defining, further,

$$L = \text{path-loss} = 10\log(4\pi R^2) \quad (dB) \qquad (3.5e)$$

and also assuming that (G_r/T) is also already expressed in dB, we simply write:

$$\frac{C}{T} = eirp - L + \frac{G}{T} - La \quad (dB/K) \qquad (3.5f)$$

where La represents any additional losses (antenna pointing error, atmospheric absorption, etc.), in dB, as well as any desired system margin, also in dB. Numerically we have:

$$L = 92.45 + 20 \log R \quad (dB) \qquad (3.6a)$$

The *power flux density, pfd* (also called the *illumination level*) is also a useful parameter, given as:

$$pfd = eirp - 163.3 - La \quad (dBW/M^2) \qquad (3.6b)$$

at the edge of coverage (worst-case).

3.2.2. Downlink

An analogous consideration for the case of the downlink can also be carried out, yielding the corresponding (downlink) C/T, C/N_o, and C/N values. The net result for the downlink can be written in terms of the downlink (C/T) -value

$$(C/T)_d = eirp - L + G/T \quad (dB/K) \qquad (3.7)$$

where now the eirp on the right of the equation (3.7) refers to the satellite downlink eirp, L is the path-loss for the downlink radiation frequency carrier EM wave, and G/T is the value of the ratio of the downlink (receive) earth station antenna to the receive (LNA) noise temperature

3.3.1. Total Link

The total link performance or signal quality can now be obtained using the relation:

$$\left(\frac{C}{T}\right)_t^{-1} = \left(\frac{C}{T}\right)_u^{-1} + \left(\frac{C}{T}\right)_d^{-1} \qquad (3.8)$$

where the suffixes u, d, and t denote the uplink, downlink, and total value, respectively. Analogous relationships hold for other ratios, C/N_o, C/N, etc.

The above parameter value for any link -- uplink or downlink, and, hence, also for the total link -- can be written as:

$$C/N_0 = C/T \text{ (dB/K)} - 10 \log(k) = C/T + 228.6 \qquad (3.9a)$$

where $- 10 \log(k) = + 228.6$ has been used. Finally,

$$C/N = C/N0 - 10 \log (B) \qquad (3.9b)$$

where B (in Hz) is the total bandwidth employed.

3.3.2. Software Implementation and Flow-Chart

Eric Johnson provided link-budget calculations by developing the public domain website.

For numerical computations, the system parameters are input appropriately. The schematic diagram (flow-chart) for this is presented in Table 3.1. Thus, to use this automated website, complete all the white-colored input boxes marked 'W', and then click on any calculator (designated buttons are marked 'C'). The pertinent results of calculation are then obtained in the corresponding output box(es) marked by the green-color (G).

Table 3.1- Schematics for Automated Link-Budget Calculations Using the Software www.satsig.net/linkbugt.htm Developed by Eric Johnson

COLOR CODE	INPUT PARAMETER/ OUTPUT RESULT	NUMERICAL VALUES	
W	Uplink Frequency (GHz)		
W	Uplink Antenna Diameter (m)		
W	Uplink Efficiency (default 65%)		
G	Uplink Antenna Transmit Gain (dBi)	G	
W	Uplink Power at the Feed (W)		
G	Uplink eirp (dBW)	G	
WC	Range (35,778-41, 679km)		
G	Uplink Path Loss (dB)	G	
COLOR CODE	INPUT PARAMETER/ OUTPUT RESULT	NUMERICAL VALUES	
C	Uplink pfd at Satellite (dBW/m^2)	G	
W	Bandwidth (Hz)		
GC	Satellite Uplink G/T (dB/K)		
G	Uplink C/N (dB)	G	
C	CLICK TO CALCULATE RESULTS		
W	Downlink Frequency (GHz)		
W	Downlink Receive Antenna Diameter (m)		
W	Downlink Receive Antenna Efficiency (e.g., default 65%)		

C	Downlink System (Antenna + LNA) Noise Temp. (K)		
G	Downlink Receive Antenna Gain (dBi)		G
G	Downlink Receive Antenna G/T (dB/K)		G
C	Downlink Satellite eirp (dBW)		
G	Downlink Path-Loss (dB)		G
G	Downlink C/N (dB)		G
C	CLICK TO CALCULATE RESULTS		
C	Uplink C/interference noise (dB)	28.0	
G	Uplink C/N (dB)		G
C	Satellite C/interference (dB)	21.0	
G	Downlink C/N (dB)		G
C	Downlink C/interference (dB)	28.0	
G	Total Link C/N (dB)		G
C	CLICK TO CALCULATE RESULTS	CLICK TO ZERO EVERYWHERE EXCEPT DEFAULTS	
	ADDITIONAL LINKS FOR SPECIAL CASES, MODES, etc.		

Note: The Link-budget Calculation with Color-coded (W-white; G-green) chart and Output Results (C) are marked above by the symbols indicated (W, G, C).

3.3. Atmospheric Propagation

Satellite communication, with the GEO or LEO satellites far above the earth's atmosphere directly affected by the weather pattern, generally suffer from propagation losses. The degree of propagation loss is similar to the nature of impulse noise or disruption in the electric circuit. However, the propagation loss critically depends on the frequency-band of the uplink and downlink EM waves. Certain frequency bands are more affected than others due to the absorption and resonance (equality of the frequency values) with respect to the quantum energy level distributions of water (present in the propagation path in the form of atmospheric moisture, and greatly increased due to precipitation in any form—rain, snow, fog, etc.), and of other atmospheric constituents (oxygen, nitrogen, carbon-dioxide, rare gases, ozone). The level of solar activity or cosmic ray shower causing a large increase in the influx of protons, electrons, and other charged particles as well as a distortion in the terrestrial magnetic field (particularly during magnetic storms caused by a solar flare, for instance) can be a major source of degradation in the fidelity and quality of the signal transmitted via a satellite. Under severe atmospheric or solar activity conditions, the signal may suffer fading. Studies on the detailed statistics of the duration and degree of fading *("fade-depth")* have also been carried out to characterize the impact of the distribution of the relevant weather (and solar activity) pattern. After all, the carrier waves carrying signal are electromagnetic waves composed of periodically varying electric and magnetic fields which are susceptible to these disturbances.

Thermal noise in the equipment usually causes random bit errors correctable by relatively simple coding. However, the above types of *'burst noise'* causes *burst errors*, characterized by bit errors in multiple consecutive or neighboring bits in the message or *code word*. Special measures—in terms of burst-error correction schemes and mathematically sophisticated coding—are required to remove or mitigate the impact of burst-noise in satellite communications channels.

The C-Band (6/4 GHz) is relatively immune from atmospheric propagation degradation, and hence was the first choice for satellite communications. But the overall spectrum of various frequency bands is of course a limited natural resource. As the C-Band frequency utilization is saturated, the need arises to consider other (generally higher) frequency bands for satellite communication.

In using higher frequency-bands, the specific propagation characteristics must be kept in mind. Different types of applications (e.g., fixed satellite services versus mobile satellite services) also dictate the choice of various frequency bands. A general overview of various frequency-bands is briefly summarized below.

(I)   Frequency Band below 2 GHz (L-Band)

The L-Band frequencies are most affected by atmospheric or ionospheric scintillations. The scintillation phenomena show diurnal and seasonal variations due to the associated variation patterns of the solar radiation causing ionization of atmospheric gases.

(II)   C-Band (6/4 GHz)

Most early satellite designs exclusively depended on the use of the C-Band for uplink (6 GHz) and downlink (4 GHz). These frequencies are least affected by normal atmospheric conditions.

(III)   Ku-Band (14/11 GHz)

As the C-Band spectra became saturated or fully utilized, satellite design moved to include the Ku-Band frequencies (14 GHz band for uplink and 11 GHz band for downlink).

(IV)   Ka-Band (30/20 GHz)

With saturation of the Ku-Band frequency spectrum, and under the need of higher capacity, the communications satellites have included the use of the Ka-Band frequencies (30 GHz band for uplink and 20 GHz for downlink.) These frequencies are most sensitive to precipitation, and the satellite design must include appropriate levels of *'rain margins'* for various applications, based on the local and regional weather patterns (e.g., average rainfall data, among other factors).

Interleaving

In the case of a *bursty* channel (that is, in the presence of burst noise causing corruption of a number of consecutive or neighboring bits), a process called *interleaving* is most helpful in the error correction process. Burst errors are caused by impulse noise in the channel, or due to deep fading produced in the data on account of precipitation (e.g., rain, fog, snow, lightening, etc., in the atmospheric propagation path or medium), occurrence of high levels of solar activity, disruption in the electric circuitry, and so on. Interleaving reduces the average number of bit errors per code word or message.

In the interleaving process, the input data is first rearranged in a rectangular or matrix form consisting of a given number of rows and columns. Thus the input bits are used to sequentially fill the first row, then the second one, and so on, until the matrix of say, m rows and n columns is completely filled. Within any column, then, the separation between successive bits is n bits. This separation is referred to as the *interleaving depth*. Coding is applied sequentially column-by-column, but the coded data is transmitted row-by-row.

The decoder also uses a matrix structure with the same numbers of rows (m) and columns (n), using an identical type of shift register. The receiver reads the decoder data column-wise, but reassembles the bits row-wise, in order to retrieve the original message.

The net result of interleaving is to minimize the impact of burst-noise in the received message. If the bit period is ρ seconds (i.e., the bit rate of $1/\rho$ bits/s), and a burst noise causes bit errors in some or all of δ consecutive bits, then, since these bits are transmitted column-wise sequentially, an impulse or burst-noise lasting for up to $n\rho$ seconds affects at the most 1 bit of the code word of length equal to 'm' bits or less. Such single bit-error in code words can be corrected relatively far more easily. This allows the reassembled (output) data to be virtually error-free. In absence of interleaving, all or most of the 'n' bits in error due to the burst-error or fade affected would have to be discarded; whereas interleaved data can be assumed to be error-free and acceptable under the above conditions. If the coding process is capable of correcting up to β errors, a burst-length up to a duration of $B_l \leq n\beta$ bits (i.e., $n\beta\rho$ seconds) could be tolerated without any corrupting effect on the output data on the receive side.

An effective or optimum interleaving design clearly depends on the characteristic pattern of the burst-noise. The prior knowledge of the maximum burst-length, for example, helps in the determination of the required interleaving depth (the number of columns in the interleaving shift register).

References:

[3] www.satsig.net/linkbugt.htm (This public-domain website also has links to calculate antenna pointing, VSAT-related calculations, etc.)

4. DIGITAL TECHNOLOGY

4.1. General

Digital Transmission involves sampling the analog (continuous in values and time) signal at sufficiently high rate, and forming a time-discrete signal using a set of quantum levels. For sampling, the signal amplitude, phase, or frequency is chosen, and the carrier wave carries the amplitude-shift-keying (ASK), phase-shift-keying (PSK), or the frequency-shift-keying (FSK) modulated signal over the link. Hybrid modulations can also be used. The digital signal is more rugged due to possibility that the degraded bit-stream can be fully regenerated accurately and the system-noise and any interference-noise can be repeatedly discarded at every stage of amplification using repeaters. This improvement of signal quality is at the expense of the requirement of a larger bandwidth, however. On the other hand, digital transmission involves lower power consumption, lighter equipment, and easier design and mass-manufacturing techniques. Replacement of analog technology by digital technology in satellite system has rapidly taken place for voice (telephony) and television (and other video) services during the recent decades. Mobile and cellular systems exclusively utilize digital technology (only), with many special advantages, including ease of mixing services (particularly text-streaming in video and TV signals), efficient mass-maneuvering, diverse inter-device conversion, smart-phone type value-added capabilities and consumer-oriented APPs (Applications), etc., at much lower cost and with enormously greater flexibility.

In this Chapter, we briefly review the basic theory of primary digital processes, focusing on various signal modulation, detection, and bit error estimation and correction schemes, without going into details of design and implementation. The important subject of multiple access is discussed in the next Chapter, and of digital coding and decoding in the following two Chapters.

4.2. Signal Modulation

4.2.1. Binary Phase Shift Keying (BPSK)

In the case of BPSK, one single bit is used to designate 0° phase (say, by the bit- value 1), and 180° (π radian) (by the bit-value 0), as symbolically illustrated in Figure 4.1.

```
                              I
                              I
Bit-value→      0_____I_____1
Signal phase→   υ = 180°       υ           υ = 0°
```

Figure 4.1. Schematic representation of the BPSK modulation scheme.

4.2.2. Quaternary Phase Shift Keying (QPSK)

The most common modulation – conventional QPSK -- involves the signal waveform of the types:

$$A \sin[w_o t + \emptyset_m(t)] = \pm \tfrac{A}{\sqrt{2}} \sin\left(w_o t + \tfrac{\pi}{4}\right) \pm \tfrac{A}{\sqrt{2}} \cos\left(w_o t + \tfrac{\pi}{4}\right) \quad (4.1)$$

with the following correspondences between the $\emptyset_m$ – value and the combination of signs ($\pm$) on the right:

sign-combination	ϕ_m (radians)	2-bit symbol
(a) +, +	0	0 0
(b) +, -	$\frac{\pi}{2}$	0 1
(c) -, +	π	1 1
(d) -, -	$\frac{3\pi}{2}$	1 0

The pertinent schemes are illustrated in Figures (4.2) and bit-phase relations are summarized below.

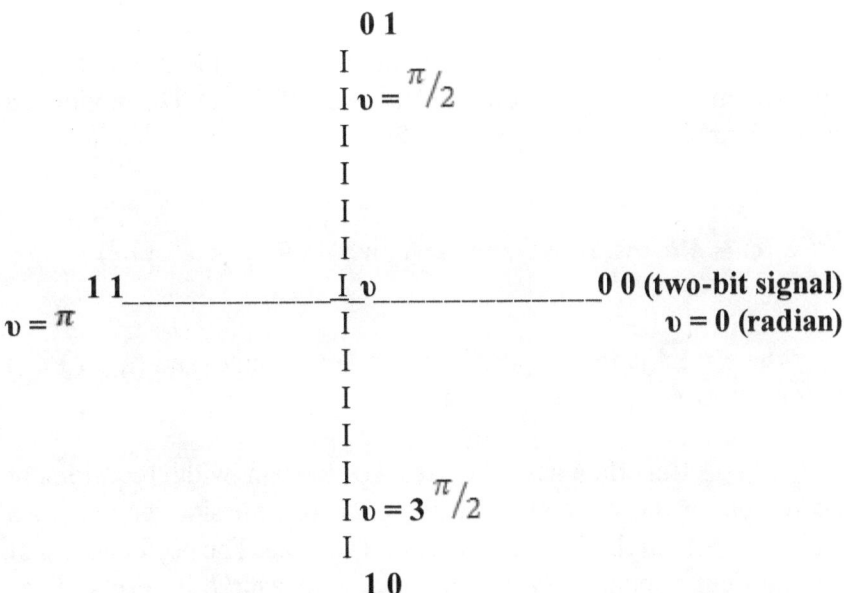

Figure 4.2. Schematic representation of QPSK modulation.

There may be further 3 types of variations in QOSK modulation scheme depending on the delay of the bit-stream, viz.

(a) The 0-QPSK -- The signal-space diagram for this type of implementation is shown in Figure 4.1.

(b) Offset QPSK (OQPSK) – In this case, the even bit stream is delayed by a half-bit interval with respect the odd 1 bit.

(c) $\pi/4$ -Shift QPSK – In this case, a phase increment of $\pi/4$ is added to each symbol.

A straightforward generalization of the QPSK is the 8Φ-PSK where a combination of three bits, viz. 0 0 0, 0 0 1, 0 1 0, 1 0 0, 0 1 1, 1 0 1, 1 1 0, 1 1 1, are used to specify 8 possible phases,

4.2.3. Quadrature Amplitude Modulation (16-QAM)

This is a hybrid scheme involving a combination of amplitude-shift-keying (ASK) and 8-phase-shift-keying (8-PSK). The modulated input signal is of the form:

$$s_i(t) = A[a_i \cos(2\pi w_c t) + b_i \sin(2\pi w_c t)], \quad 0 < t < T \quad (4.2)$$

where $\sqrt{2A}$ is the amplitude of the lowest state, and (a_i, b_i) is a pair of identifying states.

The spectral efficiency increases (i.e., the bandwidth requirement decreases for the same amount of signal-transmission) as one goes from lower- to higher-order modulation scheme. The performance of any particular modulation scheme is also measured in terms of the associated error-probability. The probability of error for a $\sqrt{M} - ary$ Pulse-Amplitude-Modulation (PAM) can be shown to be[4]:

$$P_e(\sqrt{M}) = 2\left(1 - \frac{1}{\sqrt{M}}\right) Q \left(\sqrt{\frac{3}{M-1}\frac{S_a}{N_o}}\right) \quad (4.3a)$$

where

$$S_a = 2\left(\frac{A^2 \sum_1^{\sqrt{M}} a_i^2}{\sqrt{M}}\right) \qquad (4.3b)$$

and S_a/N_o is the average signal-to-noise ratio (SNR) per symbol of an M-QAM. The probability of a symbol-error for an M-ary QAM is:

$$Pe_M = 1 - \left(1 - P_{\sqrt{M}}\right)^2 \qquad (4.4)$$

4.3. Signal Detection

Detection of digital signals is primarily performed by using the coherent detection technique. This requires a reference waveform of accurate frequency and phase, which must be properly synchronized with the use of phase-coherent carrier tracking loop for carrier recovery. Implementation techniques include the use of a phase-locked loop (PLL) oscillator or the *Costas Loop*. The synchronization problem could be avoided using differential PSK (DPSK) techniques. Simple implementation include the differential binary-phase shift-keying (DBPSK), and differential quaternary-phase shift-keying (DQPSK), entailing a symbol of one bit and two bits, respectively. Higher-order modulation are also possible and used; e.g., the 16-quaternary amplitude modulation (16-QAM).

4.4. Error Correction

There are two main error correction schemes:

(1) Automatic-Repeat Request (ARQ)
(2) Forward Error Correction (FEC)

A combination of the ARQ and FEC schemes can also be used.

The ARQ scheme is based on the retransmission of bits under error, and uses a header providing information about the source address, the destination address, and about the routing pattern. There are two types of ARQ schemes:

(a) <u>Stop-and-Wait ARQ</u> – In this scheme, after each transmission, the transmitter waits for a reply from the receiver before transmitting the next message. The messages are sent in packet form. Acknowledge (ACK) or negative acknowledge (NAK) signals serve to identify packets without and with errors, respectively.
(b) <u>Selective ARQ</u> – In this case, a message containing several words, each with its own error detection scheme, is transmitted. The word-stream is transmitted continuously, till any error in a word is detected. Then only the word with bit error(s) and all the subsequent words are retransmitted.

Expected Number of Transmissions[5]

In the case of stop-and-wait ARQ, in order to successfully transmit a message of N words, the total number of transmissions may be larger than N. The <u>expected number of transmissions</u> depends on word error rate (WER), or the probability of words-error, P_W. The expected number of transmissions, E_N, is given as:

$$E_N = \frac{1}{(1-P_W)^N} \quad (4.5)$$

Similarly, in the case of <u>Selective Retransmission with ARQ</u>, the <u>expected number of transmissions</u> with fewer than k transmission is given by:

$$E_N = \sum_{i=1}^{\infty} [1 - (1 - P_w^{i-1})^N] \quad (4.6)$$

Transmission Efficiency

Assuming k message-transmission bits are accompanied with h header bits, the total number of bits per words is:

$$b = k + h \tag{4.7}$$

The transmission efficiency, η, is defined as the ratio of actual message bits to the total number of bits including header bits and retransmission bits due to bit-error. The expressions for the transmission efficiency for various transmission schemes are as follows:

Stop-and-wait ARQ

$$\eta = \frac{Nk}{(h+Nb)E_N} \tag{4.8a}$$

where E_N is the expected number of transmissions to have the message of N words transmitted successfully (Equation 4.5).

Selective Retransmission

$$\eta = \frac{Nk}{hE_N + NbE_1} \tag{4.8b}$$

Where E_N is as given in Equation (4.6) and

$$E_1 = \frac{1}{(1-P_w)} \tag{4.8c}$$

In the above (Selective Retransmission with ARQ) case, it is assumed that both FEC and error detection coding are used.

On the other hand, in the case of Selective Retransmission with ARQ, if it is assumed that error detection parity bits are sent, but not the FEC parity bits, under the assumption of a large probability of error-free message, then the transmission efficiency can be written as:

$$\eta = \frac{Nk}{hE_N + NbE_1} \quad (4.8d)$$

where, in this case,

$$E_1 = 1 + \sum_{i=2}^{\infty}[1 - (1 - P_1 P_2^{i-2})] \quad (4.8e)$$

Probability of Undetected Error

Two types of detection-decisions may be employed:

(a) <u>Hard Detection</u> – when each transmission is characterized either 'success' or 'failure.'
(b) <u>Soft Detection</u> – when each transmission is characterized in terms of three probabilities per word, viz., P_s for success; P_d for detected error; and P_u for undetected error.

If coding is used, we can write:

$$P_s = 1 - P_w \quad (4.9a)$$

$$P_{um} = \frac{P_u}{P_s + P_u} \quad (4.9b)$$

where the second suffix ('m') on the left of Eqn. (4.9b) signifies undetected error probability per single-word message, and

$$P_u \leq P_w 2^{-m} \qquad (4.9c)$$

where m is the number of error detection parity bits. The probability of undetected error for a N-word message can be written as:

$$P_{um} = \frac{1-(1-P_w)^N}{1+(1-P_w)^N(2^m-1)} \qquad (4.9d)$$

In the case of <u>Selective Retransmission,</u> we have b parity bits per word (instead of m parity bits for the whole N-word message). Consequently, the probability of undetected error for a single-word message is:

$$P_{um} = \frac{P_w}{P_w(1-2^b)+2^b} \qquad (4.9e)$$

and the probability of undetected error for a N-word message is:

$$P_{um} = 1 - \left[1 - \frac{P_w}{P_w(1-2^b)+2^b}\right]^N \qquad (4.9f)$$

It should be noted that P_w is difficult to evaluate in the case of mobile system, since each bit in error may or may not be affected by adjacent bit error depending on the speed of the vehicle of the mobile user, making P_w a function of the vehicle speed.

4.5. Baseband Signal

The baseband signals refer to the actual basic information that is subjected to coding and modulation, and using one of the appropriate methods of multiple access, is transmitted from the transmitter to the receiver. The coded and modulated signal is carried via a radio-frequency (RF) carrier wave, and uses the satellite or terrestrial (cellular/mobile) channel (medium) for being transmitted, with necessary amplification and regeneration along the way.

In the case of the digital signals, the bit sequence involved in the message transmission system may be regenerated to mitigate or eliminate the effects of degradation due to channel noise and interference, or internal noise such as *intermodulation* noise produced in the case of multiple carriers sharing the amplifier -- due to non-linearity of the amplifier (e.g., the satellite TWTAs). This problem arises in the case of a TWTA carrying multiple carriers by sharing the bandwidth involved. This mode is called *Frequency Division Multiple Access (FDMA)*, and commonly necessitates operating the TWTA in the region of linear response, well below the *saturation point* of maximum amplification where the nonlinearity is most pronounced. This *backed-off* mode of operation in FDMA introduces some loss of efficiency.

In the case of *Time Division Multiple Access (TDMA)*, to be discussed in more detail later in this book, only one carrier occupies the TWTA at any time, so the problem of *intermodulation* noise is eliminated. This feature permits operation of the TWTA at or near the point of maximum (nonlinear) amplification (saturation), thereby enhancing the efficiency (capacity) of the TWTA. In the case of *Code Division Multiple Access (CDMA)*, on the other hand, a large number of small (in terms of bandwidth and power) carriers are used simultaneously. The transmission by other (CDMA) users using the channel simultaneously with the desired signal constitute the main source of noise. However, this noise is eliminated while retrieving the desired signal by virtue of the unique code associated with each signal.

Different Types of Baseband Signals

The type and quality (in terms of the bit error rate, R_b) of a baseband signal depends on the type of service applications. The use of *frequency modulation, frequency division multiplexed, frequency division multiple access (FM/FDM/FDMA)* mode commonly employed in earlier days has nowadays almost completely been replaced by digital modulation (e.g., BPSK, QPSK, QAM, etc.); and multiplexing and multiple access is usually also performed in the digital domain (TDM; TDMA, CDMA, etc.) in order to take advantage of digital transmissions as mentioned before.

The main type of service applications include:

(a) (Digital) Telephony
(b) (Digital) Television (DTV)
(c) Data (which is digital, of course)

New innovative applications are being added continually, including value-added services to the users. These include (digital):

(i) voicemail
(ii) texting
(iii) social and professional networking (e.g., Blogs, Facebook, Twitter, Internet, U-Tube, etc.)

Here we present the main characteristics of the various leading service applications, focusing on the digital transmission techniques. These services could be combined by proper multiplexing, a special advantage of digital transmission only. Special types of protocols (e.g., *SONET*) could be employed for large data sets to facilitate convenient multiplexing and transmissions. A large set of computer languages and methodologies are being developed at a rapid rate for special applications, as well.

4.5.1. Digital Voice (Telephony)

General

Previously, conversion to digital transmission was employed for Trunk service. The current trend is to utilize digital voice coding and multiplexing, etc., at the very source of individual voice channel (telephony), particularly for mobile and cellular services now ubiquitously available throughout the world. The mobile services may be provided via satellite(s) or via purely terrestrial (cellular) networks, or via a combination of these two methods.

Typical range of voice-encoded signals for small mobile terminals is 2.4-4.8 kbps. Various techniques of bit-reduction are used for minimizing the bit-rate. This includes reduction of the inherent redundancy in normal human speech. Reducing the bit rate significantly provides improved spectral efficiency (i.e., increases the channel capacity), though this is accomplished at the expense of increased decoder complexity and cost while degrading the speech quality. A compromise therefore is made between the spectral efficiency and system complexity and cost, in order to provide the best available speech quality at the lowest possible cost, making use of the mentioned inherent trade-off.

Two main types of speech coding are referred to as

(a) Source Coding
(b) Waveform Coding.

These two types of voice coding are described below.

4.5.1.1. Voice Coding by Source Coding

As the name implies, the source-coding performs the speech coding by quantizing and coding the essential characteristic parameters – such as the loudness or excitation level, pitch, and spectral features of the voice at the transmit end, and by performing the inverse operations at the receive end.

Thus the source-coder (vocoder) and decoder of voice basically work on the basis of extraction of the characteristic (quantized or digitized) values of the parameters of the speech and then rebuilding the analog voice output by the decoder by re-synthesizing these parameters in the inverse process by the decoder.

Bit reduction techniques as well as the vocoder (and decoder) implementation methods have rapidly undergone improvements in recent years, including the use of very large-scale integration (VLSI) technology, instrumental in reducing the cost. Bit-reduction technique is especially beneficial in satellite transmission by virtue of bandwidth-reduction, as well as in the case of mobile communication by virtue of capacity enhancement, with the use of limited available bandwidth. Typically, vocoders provide good speech quality at 4.8 kbps, with potential for further bit-reduction under continuing technological innovations, including automatic speech recognition (ASR) using Guassian Markov Model (GMM) and Deep Neural Network (DNN) model[1].

4.5.1.2. Waveform Coding

Waveform coding-decoding, as the name implies, uses the waveform of the speech signal to code and decode (reassemble) the voice, using either operation in the frequency-domain (sub-band coder) or in the time-domain, with one of the digital modulation techniques from among:

(a) Pulse Code Modulation (PCM)
(b) Differential PCM (DPCM)
(c) Adaptive DPCM (ADPCM)
(d) Delta Modulation (Δ-MOD)

4.5.1.3. PCM

A PCM modulation is commonly based on a minimum sampling rate of 2B (per second) for a signal of bandwidth B (Hz) – i.e., at the so-called *Nyquist sampling rate* -- in order to preserve the information content of the input signal. If the number of bits used to digitally code the signal is n, the total number of (quantal) levels would clearly be 2^n (e.g., $2^8 = 256$ quantized levels for 8-bit coding); so the PCM bit-rate that must be used becomes $2B2^n$ bps $(= 2^{n+1}B/1000$ kbps).

An inverse processing of the properly filtered receive signal at the receiver, with the same set of hierarchal quantum level schemes, recovers the original signal (information content).

Due to the discrete (quantization) levels and the corresponding signal processing scheme, *'quantization noise'* is produced, distorting the original (analog/voice) signal, the quantization noise increasing as the difference between the successive quantum levels increases. The signal-to-quantization noise ratio, $(S/N)_q$, is given by[6]:

$$\left(\frac{S}{N}\right)_q = 12 \left(\frac{S_r}{\Delta S}\right)^2 \qquad (4.10a)$$

where

S_r = the root-mean-square (rms) value of the input (analog) signal, and

ΔS = the difference between the successive quantal level (i.e., the quantization step)

Thus, for a given input signal (fixed rms value of the signal), the quantization noise decreases and the signal quality improves (higher value of $(S/N)_q$) as the quantization step is decreased.

In the *fixed satellite service (FSS)* with geosynchronous satellites, satellite communications most commonly use of the *quadrature* PCM (QPSK) in the QPSK/TDM/TDMA (as well as QPSK/FDM/FDMA or a hybrid) transmissions.

DPCM

The basic principle or scheme of DPCM has been given before. In the voice coding using DPCM, advantage is taken of the fact that the variation pattern (or, more specifically, the statistical variance of the *difference* between the successive levels of the signal) is smaller than the variation-pattern of the signal itself. Thus, instead of coding (by PCM technique) the signal in full, if only the differences of the successive levels of the quantized signal are coded, the number of bits required is reduced without impacting the signal quality adversely. Conversely, if the same number of bits is used to code these differences as the full signal, the signal-to-quantization noise increases by a factor α, given by:

$$\alpha = \frac{1}{2(1-R)} \qquad (4.10b)$$

where R is the auto-correlation coefficient for samples of successive signal levels. Since $0<R<1$, a higher level of correlation (i.e., higher values of R) make the advantage of using DPCM increase significantly leading to an increase in the $(S/N)_q$-value.

In general, a DPCM allows about 2-bit reduction in the required number of bits compared to regular PCM for comparable speech qualities.

ADPCM

In a DPCM, the next signal level can be estimated – or predicted – based on the preceding signal pattern (statistics or variance). If, in addition, the statistical variation-pattern, i.e., the values of the related coefficients are periodically updated (typically with 10-30 ms interval), the technique of ADPCM is realized, yielding further advantage (typically 3-bit) compared to the regular PCM technique.

Just as DPCM, ADPCM helps realization of comparable speech quality (as for PCM) with reduced bit rates, and hence with reduced bandwidth. Such reduction or bandwidth economy is beneficial for efficient utilization of the limited natural resource of bandwidth, the trade-off being somewhat increased complexity, including memory or storage requirement, for the codec involved.

Δ–MOD

In the case of Δ–MOD (pronounced as 'delta-modulation' or simply 'delta-mod'), only one single bit is used to denote the current signal, the value of this bit being taken to be 1 if the signal undergoes an increase, and 0 if the signal undergoes a decrease, compared to the preceding value.

In practice, the difference (δs) between the current value of the signal, say $s(t)$, and its predicted or reconstructed value, $s_r(t)$, obtained by integrating the output of the encoder, is evaluated so if:

$$\delta s = s(t) - s_r(t) \begin{matrix} >0, \text{ then transmit a '1'} \\ <0, \text{ then transmit a '0'} \end{matrix} \quad (4.11)$$

The resulting bit-stream is then transmitted by the transmitter coder. At the receiving end, an inverse operation based on the above convention is used by the decoder to reconstruct the signal.

When the input signal amplitude changes more rapidly than the integration time in the coder feedback loop, the reconstructed signal is distorted *("slope overload")*. On the other hand, if the input signal amplitude changes much too slowly, again, signal distortion results *("idle noise")*. These distortions can be minimized or eliminated by suitable adjustments in the integration time and/or the relative gain of the feedback loop. In particular, suitable variation in the gain of the delta- modulator (feedback loop) in a dynamic, real-time basis, as required to minimize both the slope overload and idle time distortions can provide acceptable quality speech, at about 20 kbps. While this bit-rate does not represent an advantage of the delta- modulation, it does correspond to extremely simple codec and other supporting hardware. This last variation (dynamic adjustment of the feedback loop gain) of the Δ- modulation is referred to as the *adaptive delta-modulation* (A Δ –MOD).

Broadcast and Toll-Quality Speech Signals

The general characteristics of the broadcast and toll-quality signals are provided in the following table:

TABLE 4.1
Major Characteristics of the Broadcast and Toll-Quality Voice Signals

Type	Bandwidth	Signal-to-Noise Ratio
Toll	~3.1 kHz	>30 dB
Broadcast	$\gtrsim$7 kHz	>30dB

5. MULTIPLE ACCESS TECHNIQUES

5.1. Introduction

The combined use of a common resource such as a transponder (providing tens of dBs of amplification of the received or uplink carrier wave followed by a downward frequency translation of the wave in order to avoid interference and then transmitting it as the downlink) by a multiplicity of users in order to satisfy the traffic requirements in the network in an efficient manner, is referred to as *multiple access* process.

Originally the multiple access process was accomplished by separating the individual carriers in the frequency domain. This is called *frequency division multiple access (FDMA)*. FDMA suffers from certain limitations; and subsequently *time division multiple access (TDMA)* method, based on separation of carriers in the time domain, was introduced. Yet another option is called *code division multiple access (CDMA)* based on the method of separate identification of carriers of different users with the use of unique codes. Here, we briefly outline the FDMA method and describe the TDMA and CDMA methods in more detail, since these two later methods are more common for modern digital transmission systems.

5.2. Frequency Division Multiple Access (FDMA)

FDMA is the conventional method for multiple access where the full transponder bandwidth is divided into multiple segments in the frequency domain, and each carrier is allocated one segment, on a permanent basis; i.e., all the carriers utilize the transponder simultaneously, under a time-invariant frequency band segmentation scheme.

A primary limitation of the FDMA method is that the maximum power available for the transponder in question cannot be utilized because the intermodulation products (IP) generated in the transponder limit its operating power level, as outlined below.

The degradation under the IP generation results when the following two conditions exist:

(a) The transfer characteristics of the TWTA involve nonlinearity (i.e., second and higher-order terms; and
(b) The TWTA carries more than one carrier simultaneously.

It is easy to see how such degradation occurs under the above two conditions. For instance, as a simplest case, suppose there are two carriers: A $Sin(w_1 t)$ and B $Sin(w_2 t)$ in the TWTA which has a nonlinearity represented by a quadratic term. The output of the TWTA can be then simply written as

$$O(t) = (A\ Sinw_1 t + B\ Sinw_2 t)^2$$

$$= A^2\ Sin^2 w_1 t + B^2\ Sin^2 w_2 t + 2AB\ (Sinw_1 t).(Sinw_2 t)$$

$$= \tfrac{1}{2}\ [A^2\ (1-Cos2w_1 t) + B^2\ (1-Cos2w_2 t)] + AB\ [Cos(w_1 t - w_2 t) - Cos(w_1 t + w_2 t)]$$

The $2w_1$ and $2w_2$ components could be filtered out by using a band-pass filter; however, the w_1 +/- w_2 components are likely still present in the output and constitute the intermodulation products (IP), which at once act as interference to the desired signal and also represent significant inefficiency in the system by diverting part of the output power in the useless IP. Usually second-order and third-order (and even higher-order) IP may degrade the output in such a case. In order to overcome these limitations of FDMA, TDMA technique is used.

5.3. Time Division Multiple Access (TDMA)

Time Division Multiple Access (TDMA) is a scheme under which a multiplicity of users access the resource – satellite or cellular bandwidth – in the time-domain instead of the usual frequency-domain under the conventional Frequency Domain Multiple Access (FDMA). Thus, instead of dividing the entire bandwidth into smaller segments and allocating a selected segment to a user for all times (i.e., permanently), the entire, full bandwidth is allocated to a single user for a certain length of time – typically on a microsecond scale – and then the allocation is changed to another user, and so on. Each user's allocated time duration varies according to the user's volume of traffic to be transmitted.

The set of users thus share different parts of the *TDMA Time Frame*, which is judiciously structured to periodically serve different users in rotation. Each user transmits its traffic at a very high data-rate, referred to as the *TDMA burst*, which occupies a time-length (period) as required for the amount of the traffic assigned thereto. The total bandwidth of the transponder determines its aggregate transmission capability and the associated total *frame-length* in the time-domain.

Each TDMA burst contains only one particular user's data, together with a header (*"preamble"*) providing information about the source and destination of the data for the allocated period of time, defining the pertinent *burst-length*. The TDMA burst is transmitted at a high data-rate (typically ~Mbit/sec) using special equipment (TDMA transmitter, and TDMA receiver at the receive end).

The TDMA scheme usually provides much higher efficiency of transmission, since the need to "back-off" a repeater from its saturation level, in order to minimize or eliminate the wasteful *intermodulation* products (IP) causing interference noise, is reduced or eliminated. The IP are generated due to the simultaneous presence of multiple carriers in the amplifier in the FDMA mode, on account of the nonlinear characteristics of the amplifier devices, as exemplified above. The disadvantages of the TDMA mode operation is generally higher cost of the TDMA equipment. A few different conventions have evolved for use in the TDMA (and CDMA), particularly for application in the mobile systems, as briefly outlined below.

North American TDMA System for Mobile (2G System)[7]

This section describes the North American TDMA (*NA-TDMA*) system for mobile telecommunication, forming part of the so-called 2G system. This system for digital cellular telecommunications is also called:

American Digital Cellular (ADC), or
North American Digital Cellular (NADC), or
Digital AMPS (DAMPS), or
International Standard 54 (IS-54) System.

The NA-TDMA system was designed starting in 1987 and became operational in 1990. As there was no frequency band exclusively allocated for digital cellular telecommunication, this new technology and service had to share the same frequency band as the FDMA system; i.e., coexisting with the FDMA system. Due to these factors, the first phase of NA-TDMA did not perform well. In 1994, an improved system design was introduced, replacing IS-54 by IS-136. Thus the system design was generally referred to as the *2G System*, and was subsequently superseded by more improved 3G, 4G, and 5G systems (discussed in more detail elsewhere in this book).

Transmission Design of NA-TDMA

The characteristics of one digital NA-TDMA channel, shared with FDMA (analog) channels, are as follows:

TDMA channel bandwidth	30 kHz
Number of TDMA frames per second	25
Length of each frame	40 ms

Number of time-slots in each frame	6
($\therefore$ Length of each time-slot $= \frac{40}{6} =$	6.66 ms)
Number of bits in each frame	1944
($\therefore$ Number of symbols in each frame $= \frac{1944}{2} =$	972
Number of bits in each time-slot $= \frac{1944}{6} =$	324 (i.e., 162 symbols)
Duration between successive bits	20.57 μs
Bit-rate of a radio channel	48.6 kbps
Symbol rate over a radio path	24,000 symbols/s

Two modes of operation – comprising of two frame lengths, one operated at full rate and the other at half-rate -- are used, with the following arrangements.

<u>Full-Rate</u>: Channel 1 uses time slots 1 and 4
Channel 2 uses time slots 2 and 5
Channel 3 uses time slots 3 and 6

<u>Half-Rate</u>: Channel 1 uses time slot i, ($i = 1, 2, 3, 4, 5, 6$)

<u>Frame-offset</u>: At the mobile station, the forward and reverse frame timings are offset by 206 symbols (equivalent to 1 time slot plus 44 symbols). The schematics of NA-TDMA (2G) system are illustrated in Figure (5.1).

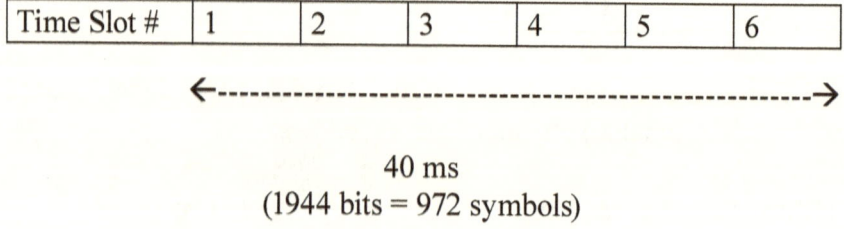

40 ms
(1944 bits = 972 symbols)

Figure 5.1. The schematics of NA-TDMA (2G) system

Power Levels: There are 11 power levels (8 in AMPS, and 3 additional ones for TDMA, with eirp for the mobile station ranging from -2 dBW to -22 dBW (for 8 AMP levels); and for the 3 TDMA levels (in the dual mode), the power levels are in the ranges of $-26 \pm 6\ dBW$, and $-34 \pm 9\ dBW$, respectively.

Modulation: NA-TDMA uses the $\pi/4$ shifted DQPSK.

System Architecture: The general system architecture of NA-TDMA is illustrated in Figure 5.2.

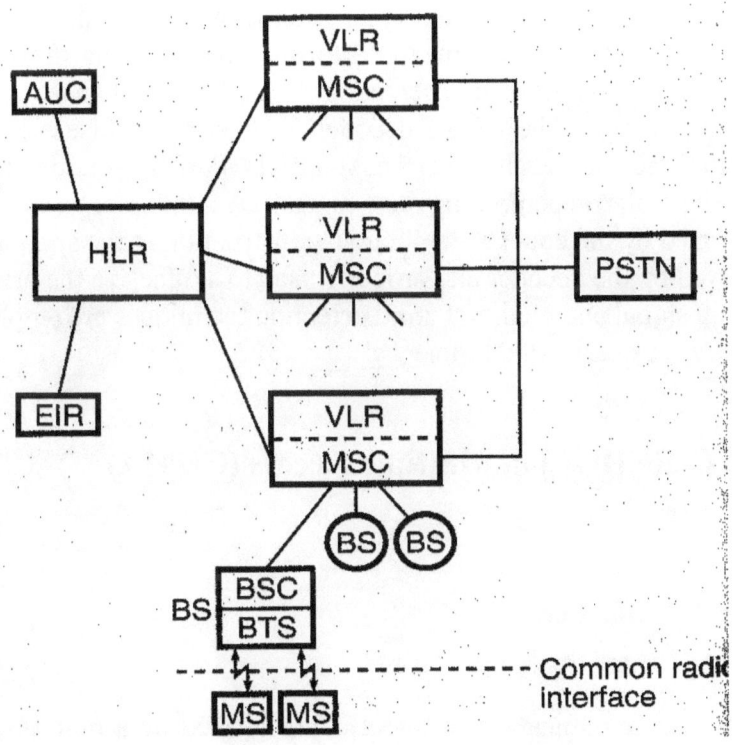

Figure 5.2. NA-TDMA system architecture
(From W. C. Y. Lee, Wireless and Cellular Telecommunications, McGraw Hill, 2006, p.132; reprinted by permission)[7]

MS: Master Station;	BS: Base Station;	HLR: Home Location Registration;
AUC: Authentication Center;	VLR: Visitor Location Registration;	EIR: Equipment Identity Register;
BSC: Base Station Controller;	MSC: Master Station Center;	BTS: Base Transceiver Station

Speech Coding Rate: The NA-TDMA uses a class of speech coding known as the *Code Excited Linear Predictive (CELP) Coding*; the specific coding used at the full rate is called *Vector-Sum Excited Linear Predictive (VSELP)* speech coding. The sampling rate of the speech coder is 7950 bps, using 20 ms long frames (containing 160 symbols). Each frame is subdivided into sub-frames 5 ms long (40 samples).

The analog speech signal emanating from the mobile station is first converted into uniform PCM format (after passing the analog signal) through a level adjuster, a band-pass filter, and an analog-to-digital converter. The VSELP decoder (a) generates a pulse excitation (Part I), and (b) speech waveform synthesis (Part II). A good quality speech results on combining the two parts. A number of parameters, generated by the coder to facilitate reconstruction of the speech, are received by the decoder and properly used to synthesize the original speech signal and quality. Error correction techniques are employed to improve quality and efficiency.

5.4. Code Division Multiple Access (CDMA)

5.4.1. Introduction

In a communications channel characterized by a high level of background noise, an effective technique is to use a rather large bandwidth to "spread" the signal in a coded manner, or to use a multiple set of frequencies over which the signal of a particular user is transmitted, changing the carrier frequency in a systematic manner by "hopping" over a large overall bandwidth. Thus it is possible to transmit and receive a weak signal by utilizing certain code available to both the transmit and receive (pair of communicating) earth-stations, in the presence of relatively large amount of background noise. In fact, noise may include the signals of other pairs of communicating users (earth stations or even hand-held "terminals"), each pair using its own unique code. This Code Division Multiple Access (CDMA) thus permits successful communication where the high noise level and mutual interference would normally make it rather difficult or even impossible to attain a workable link using the FDMA or TDMA techniques. This is achieved at the expense of a high order of bandwidth, assuming the availability of the excess amount of bandwidth for CDMA. CDMA technique is a good option for mobile and military communication where use of the unique code by a transmitter and its intended receiver permits their communication, but to the other receiver the background noise plus interference appears simply as Additive White Gaussian Noise (AWGN) in the channel, and thereby offers security against unintended (enemy) receiver. In the case of mobile communication, similarly, use of portable or hand-held terminals is necessary, which would otherwise yield a very small signal-to-noise ratio (S/N); and yet communication between interlinked users is rendered possible through the use of CDMA. A large number of simultaneous links in the network of users also becomes allowable, a characteristic requirement of mobile communications networks.

Two Classes of CDMA

From the above introduction, it follows that CDMA method can be implemented by using one of the following two categories of techniques:

(a) "Spread-Spectrum" CDMA (SS/CDMA)
(b) "Frequency-Hopping" CDMA (FH/CDMA)

These two techniques are briefly discussed below.

5.4.2. Spread-Spectrum CDMA (SS-CDMA)

A block diagram of the essential elements of a spread-spectrum CDMA implementation scheme is shown in Figure 5.3.

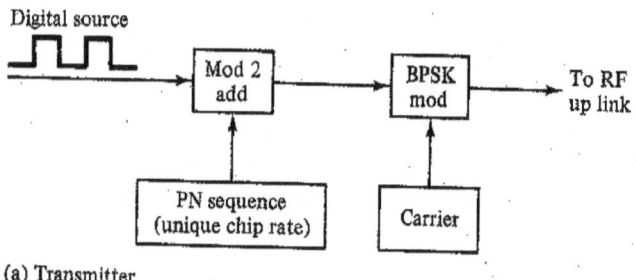

(a) Transmitter

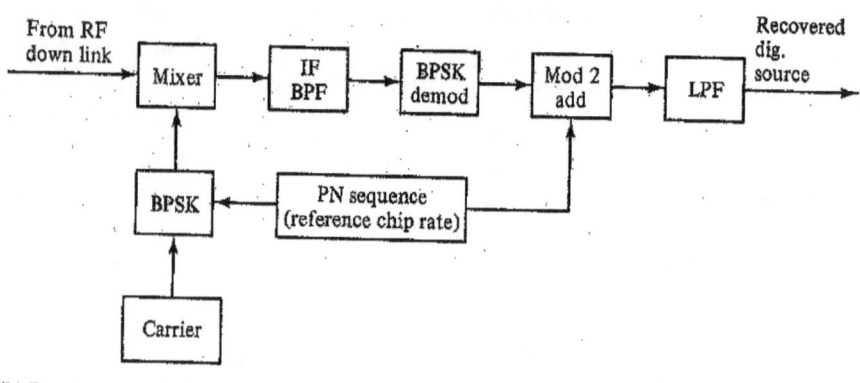

(b) Receiver (correlation detector)

Figure 5.3. Schematic Diagram of the SS/CDMA System (a) Transmitter (Top); (b) Receiver (Bottom) [from Reference 8, reprinted by permission from Pearson Education, Prentice Hall].

Thus, the CDMA transmitter (Figure 5.3a), transmits its message after coding it with a pseudo-random sequence (PN-sequence), using a large bandwidth over which the message bits and the PN-sequence bits are "spread." The receiver, using an identical PN-sequence, is able to retrieve the original signal and message. Of course the actual implementation makes use of intermediate frequency (IF) band-pass filter (BPF) and low-pass filter (LPF) and other electronics for proper processing of the signal.

The processing gain (G) of the spread-spectrum CDMA system is defined as

$$G = \frac{B_s}{R_b} \qquad (5.1a)$$

where B_s is the RF bandwidth and R_b is the bit-rate. It is also useful to define the spread-factor, F_s, as

$$F_s = \frac{T_b}{T_s} \qquad (5.1b)$$

where T_b and T_s are the bit-duration and symbol duration, respectively, and the following equality holds:

$$E_s R_s = E_b R_b \qquad (5.1c)$$

where E is the energy per bit and R is the rate, and the suffixes s and b refer to a symbol and an information bit, respectively.

Now the total noise in the channel with N CDMA users with equal power level, out of which one user (transmitter) is the desired one while $(N-1)$ CDMA users' transmission simply act as AWGN, the total noise level could be written as the sum of the actual noise in the desired user channel, plus the power of the rest $(N-1)$ users acting as background noise. Denoting the noise power desnity of a single (desired) user as N_o, therefore the total noise density N'_o, in the channel can be expressed as the sum[5]:

$$N'_o = N_o + (N-1) \times \frac{R_s E_s}{B_s}$$

The above relation could be used to determine the number of CDMA users allowed in the channel:

$$N = 1 + (N'_0 - N_o) \times \frac{B_s}{R_s E_s}$$

i.e.,

$$N \cong N'_o \left(1 - \frac{N_o}{N'_o}\right) \frac{B_s}{E_b R_b} \qquad (5.2a)$$

where we have neglected 1 (<<N) on the right of Equation (5.2a) and used Equation (5.1c). The above result can be further rewritten in the form (using Equation 5.1a)

$$N \cong \frac{G}{\left(\frac{E_b}{N'_o}\right)} \left(1 - \frac{N_o}{N'_o}\right)$$

i.e.,

$$N = \frac{G}{(E_b/N'_o)} \left[\left(\frac{E_b}{N'_o}\right)^{-1} - \left(\frac{E_b}{N_o}\right)^{-1}\right] \qquad (5.2b)$$

or

$$N = \frac{G}{(E_b/N'_o)} \left[1 - \frac{C/N'_o}{C/N_o}\right] \qquad (5.2c)$$

With usually available values E_b/N'_o, etc., several hundred users could be accommodated in a CDMA channel.

PN Sequence

The pseudo-random sequence used for the spread-spectrum CDMA can be generated using a shift-register with about equal numbers of '1' and '0,' so that the sequence generated appears to be random.

If the shift register has s stages, the maximum possible length of the code generated is $(2^s - 1)$, which is referred to as *maximum-length linear shift register sequence*, and contains $(2^s - 1)$ ones and $(2^{s-1} - 1)$ zeros. The maximum number, N_m, of maximum-length linear codes of this type of shift register, which equals the maximum number of users in the CDMA network, is given by[9]:

$$N_m = [E(2^s - 1)]/s \qquad (5.3)$$

where $E(2^s - 1)$ is a Euler Number. The value of N_m rises rapidly as s increases. Examples of the maximum number of users for different values of s are provided below (Table 5.1).

Table 5.1
The Rapid Rise of Nm with Increasing s-Value
[Examples of the Maximum Number (N_m) of users in a CDMA Network corresponding to Different Values of the Number of Stages (s) of the Shift Register.]

s	N_m
14	756
15	1800
16	2048

Power Spectral Density of the PN Sequence

The power spectral density of a PN-sequence is similar in shape to a $\sin^2 wt/(wt)^2$ function. An exact expression of the spectral density function, $f_s(w)$, is given as[9]:

$$f_s(w) = \left(\frac{A^2}{s^2-A^2}\right)(s+1)\left[\frac{\sin^2(wT_p/2)}{(wT_p/2)^2}\right]\sum_{n=-\infty}^{\infty} f_s\left(w - \frac{2\pi n}{T_s}\right)$$

$; n \neq 0)$

(5.4a)

where

s = Number of stages in the shift register
  = Number of bits in the PN-sequence
A = pulse-height (in volts)
T_p = bit-period
T_s = period for a complete PN sequence
  = $(T_p s)$

Recall that $f_s(w)$ acts like an impulse function. Since the number of ones and zeros in the sequence always differ by 1, the sequence is not truly random. This causes a very small DC term, representing the peak power at the center-frequency. This center-frequency power can be shown to be given by:

$$f_m = \frac{A^2}{s^2} f_s(w)$$

(5.4b)

The code rate is equal to the difference between the first null (where $f_s(w) = 0$) and the center frequency. An increase in the bit-rate makes the required CDMA bandwidth also increase. Typically, the RF bandwidth used is 100-1000 times larger than that required for the information bits. In order to minimize the cross-correlation between the PN-sequence codes, the number of the maximum length linear codes may be restricted. A number of codes providing a very low cross-correlation are available in the literature[10].

5.4.3. Frequency Hopping CDMA (FH-CDMA)

In the frequency hopping CDMA (FH/CDMA) system, a set of carrier frequencies are used to transmit the message in the presence of noise; with the choice of the carrier frequencies forming a random sequence, generated by an s-stage shift register, as in the case of a spread-spectrum CDMA system. This frequency hopping process thus 'randomly' selects a carrier from a set of carriers where consecutive carriers are a fixed difference, Δf, apart. These various frequencies are determined or generated by using a frequency synthesizer.

The transmitting terminal thus 'hops' over a discrete set of operating carrier frequencies, the frequency value being selected from the set at a particular step by following the (pseudo-) random sequence with the help of the shift register.

The receiver follows an identical pattern of frequency hopping, with the help of an identical shift register, which generates the same code as followed by the transmitter. The transmitter and receiver are properly synchronized. Unintended receivers do not have the same code and hence cannot follow the same sequence of the carrier frequencies, and therefore are unable to track or decode the message. With a large number of users in the network, the transmission of all the other transmitters appear as a background characterized as additive white Gaussian noise (AWGN), but the intended receiver is able to retrieve the message by using the same 'randomly' coded carrier frequencies in synchronism with the transmitter. The transmitter and receiver functions are represented in block-diagram form in Figure 5.4.

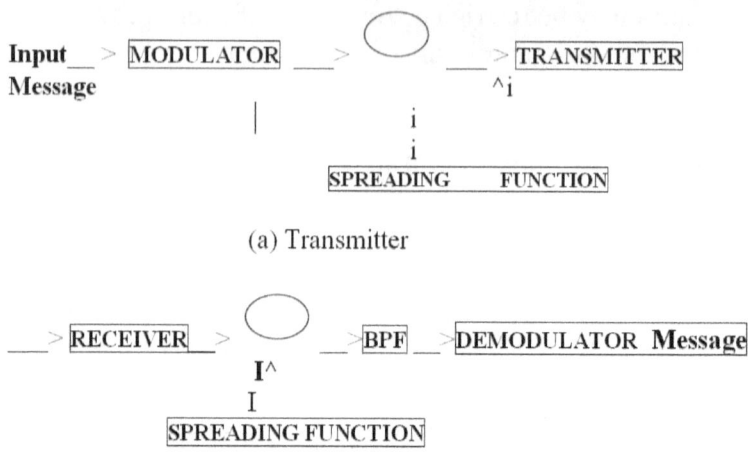

(a) Transmitter

(b) Receiver

Figure 5.4. Frequency hopped CDMA scheme, (a) Transmitter, (b) Receiver.

For an s-stage shift register, the number of maximum-length linear codes is $(2^s - 1)$, as in the case of the P-N sequence spread-spectrum CDMA described above. With a separation of the adjacent coded carrier frequencies equal to Δf, therefore, the total CDMA bandwidth, B_{FH}, is simply given as:

$$B_{FH} = (2^s - 1)\, \Delta f \qquad (5.5a)$$

The processing gain in this case, G_p, can be written as the ratio of the total bandwidth with FH/CDMA to the frequency interval Δf, i.e.,

$$G_p = \frac{B_{FH}}{\Delta f} = 2^s - 1 \qquad (5.5b)$$

A code rate much smaller than that of a PN sequence CDMA is useable in a FH/CDMA system.

References:

[4] William Y. S. Lee, Wireless and Cellular Telecommunications, McGraw Hill, p. 95, 2006.
[5] William Y. S. Lee, Wireless and Cellular Telecommunications, McGraw Hill, p. 101, 2006.
[6] M. Richharia, Satellite Communications Systems, McGraw Hills, 1999, p. 250.
[7] William Y. S. Lee, Wireless and Cellular Telecommunications, McGraw Hill, p. 131, 2006.
[8] W. L. Pritchard et al, Satellite Communication Systems Engineering, Prentice Hall, p. 389, 1993.
[9] M. Richharia, Satellite Communications Systems, McGraw Hills, 1999, p. 391, 1999.
[10] D.V. Sarvate and M.B. Pursley, Cross Correlation Properties of Pseudo-Random and Related Sequences, Proc-IEEE, Vol. 68, p. 593-619, 1980.

6. CODING – I

6.1. Introduction

The fidelity of the digital signals may be enhanced by adding to it redundant bits in a systematic way in order to detect and possibly correct bit errors. This process is called coding. Decoding is the inverse process of recovering the actual signal by removing the redundancy imbedded in the coded signal. Thus the coder at the transmitter, together with the decoder at the receive end, perform the coding-decoding process to recover the message from the corrupted RF signal. A variety of coding theory and coding methods have been developed for various applications. Here a few selected coding techniques particularly suitable for satellite communications and mobile satellite communications, usually referred to as fixed satellite services (FSS) and mobile satellite services (MSS), respectively, are briefly described, starting with the basics of information theory closely related to error estimation and correction.

A model of the signaling system with coding/decoding is provided in Figure 6.1.

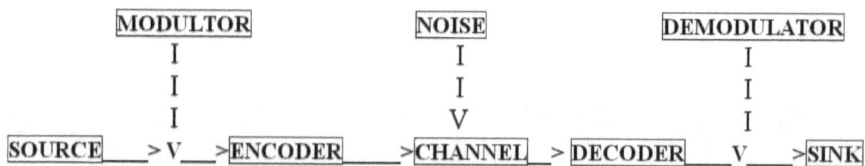

Figure 6.1. Block diagram representing a model of signaling system.

6.2. Basic of Information Theory

Shanon[11] developed the mathematical information theory, according to which the *amount of information* in a message is defined as:

$$I = \log_2 \left(\frac{1}{P_k}\right) \qquad (6.1a)$$

where P_k is the probability of occurrence of the kth message. Thus a less probable message (P_k smaller) carries more information. As an extreme, if the message has the highest possible probability ($P_k = 1$), there is no information ($I = 0$) associated with it.

In the case where a set of M messages (i.e., $k = 1, 2, ..., M$) are combined, the total information I_t is the sum of the M information for the message set:

$$I_t = \sum_{k=1}^{M} I_k = \sum_{k=1}^{M} w_k \log_2 \frac{1}{P_k} \qquad (6.1b)$$

where w_k is the weight of the kth message. If the total number of message sequences is L, then:

$$w_k = LP_k$$

Hence,

$$I_t = \sum_{k=1}^{M} LP_k \log_2 \frac{1}{P_k} \qquad (6.1c)$$

The entropy, H, associated with this set of message sequences is defined as:

$$H = \frac{I_t}{L} = \sum_{k=1}^{M} P_k \log_2 \frac{1}{P_k} \qquad (6.2a)$$

The entropy function provides a lower bound on L. If the number of messages generated per second is N, then the average information rate (R) of the source is given as:

$$R = NH \text{ bits/s} \qquad (6.2b)$$

Now, it can be shown that the entropy is a maximum when the probabilities of occurrence of all the M messages is equal (say ρ), i.e., $P_k = \rho$ (constant).

Then,

$$\sum_{k=1}^{M} P_k = \sum_{k=1}^{M} \rho = \rho M$$

But,

$$\sum_{k=1}^{M} P_k = 1$$

Hence,

$$\rho M = 1$$

or

$$\rho = \frac{1}{M}$$

Consequently, in this case, putting $P_k = \rho$ for all k in equation (6.1c)

$$I_t = \sum_{k=1}^{M} \rho L \log_2 \frac{1}{\rho} = \rho L \sum_{k=1}^{M} \log_2 M$$

$$= \rho L M \log_2 M = L \log_2 M$$

Hence,

$$H = I_t/L = \log_2 M \qquad (6.2c)$$

When the messages are not equally likely, the average information per bit reduces.

6.3. Shanon's Theorem

Shanon's Theorem is a fundamental principle of communication theory. It states that:

"When M (>1) equally likely messages are transmitted at an information bit-rate of R bits/s through a channel of capacity C (>R), then it is possible to transmit information with any desired accuracy (i.e., with any desired level of bit-error) by appropriately coding the messages."

The capacity (C) of a channel limited by Gaussian noise is given by Shanon-Hartley theorem:

$$C = B \log_2 \left(1 + \frac{S}{N}\right) \text{ bits/s} \qquad (6.3a)$$

where

B = channel bandwidth

$\frac{S}{N}$ = signal-to-noise ratio at the input of the receiver

This theorem (Equation 6.3a) is only valid for Gaussian noise (i.e., dominant thermal noise in the radio links), but it provides the lower limit (performance bound) for non-Gaussian channels.

6.4. Bandwidth-Power Tradeoff

The above theorem (Equation 6.3a) provides the basis of inherent trade-off between power (S/N) and bandwidth in any communications channel including the satellite link. Note that with sufficient increase of power (with fixed B), $\frac{S}{N} \to \infty$ and hence $C \to \infty$. However, increasing the bandwidth also increases the channel noise (N), decreasing the S/N-value. The limiting capacity value for $B \to \infty$ is then given by:

$$C_o = 1.44 \frac{S}{N_o} \qquad (6.3c)$$

where N_o is the noise power density in the channel.

The above limit in the capacity can be achieved by providing suitable coding in the signal. Now the basic principle, parameters, and types of coding suitable for satellite channels are discussed below.

Error Detecting Codes

A long message can be divided into blocks of $(n-1)$ digits and one digit added to the block which helps to detect an error within the message block. Typically, this extra coding bit represents the parity (sum of the bits modulo 2 being 0 or 1), representing even or odd parity, respectively. In the block, now we have $(n-1) + 1 = n$ bits, of which only $(n-1)$ bits contain the message. The ratio:

$$\frac{Total\ number\ of\ bits\ in\ the\ block}{Number\ of\ message\ bits} = \frac{n}{n-1} = 1 + \frac{1}{n-1}$$

(6.4)

is called the *redundancy*, the *excess redundancy* being $1/(n-1)$.

More generally, in a coded block of n bits, if the number of message bits is k (i.e., there are $n-k$ coding bits), then we refer to the system to have a (n, k) coding and the ratio $\frac{n}{k} = 1 + \frac{n-k}{k}$ is the redundancy.

For long messages, low redundancy is achievable. However, for better reliability (low bit error rate), shorter messages are preferable. Thus a trade-off in the block-length and redundancy (or bit error rate) exists and coding design must take into account this inherent trade-off.

6.5. The Effect of White Noise

If

(a) the probability of an error in the presence of noise is the same of each bit-position, and
(b) the occurrence of bit error in any position is independent of occurrence of bit error in any other position (i.e., there is no correlation between bit errors in different positions), then the noise is referred to as the *white noise*. Often, in absence of knowledge about the nature of noise in a specific situation, generally the noise is assumed to be white noise.

Let the message be n bits long under a white noise environment, the probability of occurrence of bit error in any of the n positions being p. The probability that the message is transmitted without error is $(1-p)^n$.

The probability of a single bit error in the message in any one of the n positions (which can be chosen in n ways), i.e., the probability that $(n-1)$ bit positions are error-free, is given by (the probability p of a single bit error multiplied by the probability of $(n-1)$ bits having no error, multiplied by the number n of ways in which one bit can be chosen out of n bits)

$$np(1-p)^{n-1} \tag{6.5a}$$

Similarly, the probability of 2 bit errors is

$$\frac{n(n-1)}{2}p^2(1-p)^{n-2} \tag{6.5b}$$

and so on. By extension, the probability of k bit errors is given by the $(k+1)^{\text{th}}$ term in the binomial expansion:

$$1 = [(1-p)+p]^n$$

$$= (1-p)^n + np(1-p)^{n-1} + \frac{n(n-1)}{2}p^2(1-p)^{n-2} + \cdots + p^n$$

$$\uparrow \qquad \uparrow \qquad \uparrow \qquad \qquad \uparrow$$
0 bit error 1 bit error 2 bit errors n bit errors

$$= \sum_{k=0}^{n} C(n,k)p^k(1-p)^{n-k} \tag{6.5c}$$

where $C(n,k)$ is the coefficient of the kth term in the binomial expansion. By adding the *right hand side* of equation (6.5c) to the binomial expansion:

$$[(1-\rho)-\rho]^n = \sum_{k=0}^{n}(-1)^k C(n,k)\rho^k (1-\rho)^{n-k}$$

(6.6a)

and dividing the sum by 2, we get the probability of an *even* number of bit errors

$(k = 0, 2, 4, 6, \ldots ; \text{i.e.,} k = 2m \text{ errors},\ m = 0, 1, 2, 3, \ldots)$

$$P_e = \frac{1+(1-2\rho)^n}{2} = \sum_{m=0}^{\{n/2\}} C(n,2m)\rho^{2m}(1-\rho)^{n-2m}$$

(6.6b)

where the $\{\}$ bracket means the greatest integer within $\{\}$.

The probability of an odd number of bit errors is then obviously given by

$$P_o = 1 - P_e \qquad (6.6c)$$

Since the first term of equation (6.6b) represents probability of no error, the probability of an undetected error P_u is obtained by dropping this (first) term; i.e.,

$$P_u = \sum_{m=1}^{\{n/2\}} C(n,2m)\rho^{2m}(1-\rho)^{n-2m} \qquad (6.6d)$$

6.6. Modular Arithmetic

As is well known, the mod (i.e., modulo) 2 arithmetic that simple binary parity checks use, is the same as logical addition (exclusive OR, XOR) and follows the following rules:

$$0 + 0 = 0$$
$$0 + 1 = 1$$
$$1 + 0 = 1$$
$$1 + 1 = 0$$

For multiplication, we have the rules (logical AND)

$$0 \times 0 = 0$$
$$0 \times 1 = 0$$
$$1 \times 0 = 0$$
$$1 \times 1 = 1$$

6.7. Shift Register

The coding process often makes use of *shift registers* and modulo 2 adder. These devices are based on linear digital circuits which together act as linear code generators and which, in turn, are based on the linear algebraic operations mod 2.

A shift register consists of a cascaded series of digital flip-flop comprising of a number of stages. Digital bit stream is fed at the input end of the shift register, such that, as it enters the first stage, the bits present at every stage shift to the next higher stage, either synchronously (by synchronizing the shifting process with an accurate and reliable clock), or asynchronously (independent of one another). The resultant output represents the coded data, its structure depending on the structure of the shift register and on specific shifting algorithm employed for the implementation of the shift register (i.e., the number of stages, number of mod 2 adders, their relative positions and the feedback arrangement, if any), as illustrated in the schematic diagram below (Figure 6.1).

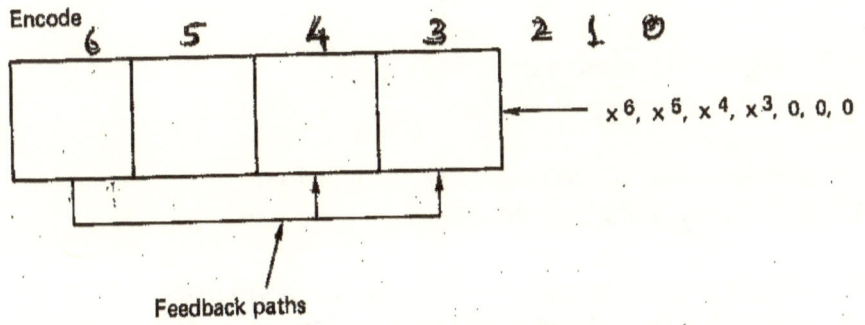

Figure 6.1. the concept of shift register.

6.8. Hamming Distance[12]

If we consider two blocks of data of equal code or *word length*, and count the number of positions where the bits are different, then this number is said to represent the *'distance'* between the two blocks or code words.

For a set of N blocks of data or code words, each n bits-long, say,

Block i: 1 1 0 1 0 1 1 1 ... (n bits)
Block j: 1 0 0 1 1 0 1 0 ... (n bits)
$$i, j = 1, 2, 3, \ldots, N; (i \neq j)$$

and denote the distance between the two blocks i and j as d_{ij}, then the minimum value of the full set $\{d_{ij}\}$ is called the Hamming distance, d_H, i.e.,

$$\text{Hamming distance} = d_H = min\{d_{ij}\}; \; i, j = 1, 2, \ldots, N; (i \neq j)$$

where the abbreviation *min* means the minimum value (integer number), and {} represents the full set of the N blocks.

The Hamming distance is a useful concept in block coding.

6.9. Algebraic Coding Theory

Any sequence of the binary digits 0 and 1 can be represented in terms of a polynomial, as follows:

The first bit from right is taken as the coefficient of $x^0 = 1$; the second bit from the right is taken as the coefficient of $x^1 = x$; the third bit from the right is taken as the coefficient of x^2; and so on. Thus, the kth bit from right is taken as the coefficient of x^{k-1} ($k = 1, 2, 3, ...$). This process yields a polynomial in x of degree $n - 1$ for a bit sequence of n bits, each term of which has only the coefficient of 1.

For instance, let us consider an arbitrary bit sequence of 7 bits, e.g.

1 0 1 0 0 1 1

Hence, putting all the terms as an addition, we get a polynomial in x of degree 6 (= 7 − 1):

$$1 \times x^6 + 0 \times x^5 + 1 \times x^4 + 0 \times x^3 + 0 \times x^2 + 1 \times x^1 + 1 \times x^0$$
$$= x^6 + x^4 + x + 1$$

The polynomial can thus be seen as the *generator* function representation of the given bit sequence.

The addition of any two polynomials of a given degree, following the mod 2 addition rule (0 + 0 = 0; 1 + 0 = 1; 0 + 1 = 1; 1 + 1 = 0) will yield another polynomial of the same or a lower degree. The addition is associative, i.e.,

$$P_1(x) + [P_2(x) + P_3(x)] = [P_1(x) + P_2(x)] + P_3(x)$$

There exists a zero element polynomial $P_0(x)$ corresponding to all bits being zero (viz., 0 0 0 0 0 0) so that, for a polynomial of degree k,

$$P_0(x) = 0 \times x^{k-1} + 0 \times x^{k-2} + 0 \times x^{k-3} + \cdots + 0 \times x^0 = 0$$

Also, each non-zero polynomial is its own additive inverse; i.e., denoting the inverse of the polynomial $P_1(x)$ by $P_1^{-1}(x)$, we note that, for $P_1^{-1}(x) = P_1(x)$,

$$P_1(x) + P_1^{-1}(x) = P_1(x) + P_1(x) = (1+1)P_1(x)$$
$$= 0 \times P_1(x) = 0 \quad \text{(the zero element)}$$

Recall that the sum of any two different polynomials is another polynomial of the same or lower degree. It follows that the set of all polynomials of a given degree (n) forms a *group*.

Also, it is clear that the product of two polynomials, say, $P_1(x)$ of degree n_1 and $P_2(x)$ of degree n_2, is also a polynomial, $P_3(x)$

$$P_3(x) = P_1(x) \times P_2(x)$$

and the degree of this product polynomial is the sum, $(n_1 + n_2)$, of the degrees of the two polynomials.

Monic Polynomial

A polynomial of degree n, with its highest order term (x^{n-1}) having a coefficient 1 (and not 0) is called a *monic polynomial*. (Note that for any x, $x^0 = 1$).

Prime Polynomial

A *prime polynomial* is a polynomial which *cannot* be represented as the product of two non-trivial polynomials. Thus a prime polynomial is a monic polynomial which *cannot* be factored into a product of two lower-order polynomials.

Just as a large prime number (which cannot be factored into two integers; i.e., cannot be expressed as a product of two integers) is an important subject in Number Theory, prime polynomials are important for coding theory and applications.

We define arithmetic modulo n when we resct $n \to 0$ for manipulation (addition, subtraction, multiplication, division), so only the numbers $0, 1, 2, \ldots (n - 1)$ appear in various operations.

The purpose of modular arithmetic, modulo 2, viz., $(0 + 0 = 0;\ 1 + 0 = 1;\ 0 + 1 = 1;\ 1 + 1 = 0;\ 0 \times 0 = 0;\ 1 \times 0 = 0 \times 1 = 0;\ 1 \times 1 = 1)$ is to be able to keep all the numbers within the range (0,1). This feature is then useful for application of coding which can be applied for computer applications. Various decimal numbers can then be expressed in terms of binary coding, using binary digits (bits).

One can conceive of modulo 4, modulo 8, etc. arithmetic. Below, the results of such arithmetic are represented, together with the corresponding decimal numbers.

Modulo 2 Arithmetic

Decimal	Code (1 bit)
0	0
1	1

[i.e., only 0, $(1 = 2 - 1 = 2^1 - 1)$, appear in arithmetical operations]

Modulo 4 Arithmetic

Decimal → Binary (2 bits)

$$0 = 0 \quad \rightarrow 00 = 0 \times 2^1 + 0 \times 2^0$$
$$0 + 1 = 1 \rightarrow 01 = 0 \times 2^1 + 1 \times 2^0$$
$$1 + 1 = 2 \rightarrow 10 = 1 \times 2^1 + 0 \times 2^0$$
$$2 + 1 = 3 \rightarrow 11 = 1 \times 2^1 + 1 \times 2^0$$

[i.e., only 0, 1, 2, $(3 = 4 - 1 = 2^2 - 1)$, appear in arithmetical operations]

Module 8 Arithmetic

Decimal → Octal (3 bits)

$$0 \rightarrow 000 = 0 \times 2^2 + 0 \times 2^1 + 0 \times 2^0$$
$$1 \rightarrow 001 = 0 \times 2^2 + 0 \times 2^1 + 1 \times 2^0$$
$$2 \rightarrow 010 = 0 \times 2^2 + 1 \times 2^1 + 1 \times 2^0$$
$$3 \rightarrow 011 = 0 \times 2^2 + 1 \times 2^1 + 1 \times 2^0$$
$$4 \rightarrow 100 = 1 \times 2^2 + 0 \times 2^1 + 0 \times 2^0$$

$5 \rightarrow 101 = 1 \times 2^2 + 0 \times 2^1 + 1 \times 2^0$

$6 \rightarrow 110 = 1 \times 2^2 + 1 \times 2^1 + 0 \times 2^0$

$7 \rightarrow 111 = 1 \times 2^2 + 1 \times 2^1 + 1 \times 2^0$

[i.e., only numbers 0, 1, 2, 3, 4, 5, 6, ($7 = 8 - 1 = 2^3 - 1$), appear in all operations]

Modulo 16 Arithmetic

Similarly, writing powers of 2 in reverse order, for the case of modulo 16 arithmetic, we have:

Decimal

(4 bits)

0 $0000 = 0 \times 2^0 + 0 \times 2^1 + 0 \times 2^2 + 0 \times 2^3$

1 $0001 = 1 \times 2^0 + 0 \times 2^1 + 0 \times 2^2 + 0 \times 2^3$

2 $0010 = 0 \times 2^0 + 1 \times 2^1 + 0 \times 2^2 + 0 \times 2^3$

3 $0011 = 1 \times 2^0 + 1 \times 2^1 + 0 \times 2^2 + 0 \times 2^3$

4 $0100 = 0 \times 2^0 + 0 \times 2^1 + 1 \times 2^2 + 0 \times 2^3$

5 $0101 = 1 \times 2^0 + 0 \times 2^1 + 1 \times 2^2 + 0 \times 2^3$

6 $0110 = 0 \times 2^0 + 1 \times 2^1 + 1 \times 2^2 + 0 \times 2^3$

7 $0111 = 1 \times 2^0 + 1 \times 2^1 + 1 \times 2^2 + 0 \times 2^3$

8 $1000 = 0 \times 2^0 + 0 \times 2^1 + 0 \times 2^2 + 1 \times 2^3$

9 $1001 = 1 \times 2^0 + 0 \times 2^1 + 0 \times 2^2 + 1 \times 2^3$

10 $1010 = 0 \times 2^0 + 1 \times 2^1 + 0 \times 2^2 + 1 \times 2^3$

11   $1011 = 1 \times 2^0 + 1 \times 2^1 + 0 \times 2^2 + 1 \times 2^3$

12   $1100 = 0 \times 2^0 + 0 \times 2^1 + 1 \times 2^2 + 1 \times 2^3$

13   $1101 = 1 \times 2^0 + 0 \times 2^1 + 1 \times 2^2 + 1 \times 2^3$

14   $1110 = 0 \times 2^0 + 1 \times 2^1 + 1 \times 2^2 + 1 \times 2^3$

15   $1111 = 1 \times 2^0 + 1 \times 2^1 + 1 \times 2^2 + 1 \times 2^3$

[i.e., only numbers 0, 1, 2, 3, ..., ($15 = 16 - 1 = 2^4 - 1$), appear in all operations]

A set of 8 binary digits are used in the 7-bit ASCII code which provides representation of the English alphabet (A, B, ... Z), numeric (0, 1, 2, ...), and other symbols. An 8th bit is commonly used as a parity check.

Thus, as the above examples show, the purposes of modular arithmetic is to ensure that all the numbers that can appear will remain within a given range (which is equal to 1 less than the value 2^n, where n is the order, i.e., the number of digits in the binary representation that appear is equal to $2^n - 1$).

6.10. Polynomial Coding

The method of numeric binary coding can be extended to the realm of polynomials in x as defined above, where any sequence of binary digits (bits) 0 and 1 can be converted into a polynomial of x with various coefficients of x being the bit in the corresponding position. Thus for the binary bit sequence

1   1   0   0 ... 0   1   (n bits),

the polynomial representation becomes

$$1 \times x^{n-1} + 1 \times x^{n-2} + 0 \times x^{n-3} + 0 \times x^{n-4} + \cdots 0 \times x^1 + 1 \times x^0$$
$$= x^{n-1} + x^{n-2} + \cdots + 1$$

For example, with 5 bits sample

1 0 1 0 1

the associated polynomial is

$$1 \times x^4 + 0 \times x^3 + 1 \times x^2 + 0 \times x^1 + 1 \times x^0 = x^4 + 0 + x^2 + 0 + 1$$
$$= x^4 + x^2 + 1$$

A product of two monic polynomials of order n and m is clearly another monic polynomial of order $(n + m)$, as mentioned above:

$$(x^n + x^{n-1} + \cdots 1) \times (x^m + \cdots + 1)$$
$$= x^{n+m} + \cdots + 1$$

The two component polynomials are called the factors of the composite polynomial of order $(n + m)$.

According to the definition stated above, a prime polynomial is a ***monic*** polynomial which cannot be factored into two component polynomials of lower order. Prime polynomials are of special interest and importance in algebraic coding theory, as discussed below.

First, we obtain the sets of prime polynomials of various degrees (maximum power of x, which by definition must have coefficient 1, and *not* 0)

Degree 0: For degree 0, the only term or polynomial is

$$P_o = x^0 = 1$$

This trivial polynomial actually corresponds to the arithmetic number 1.

Degree 1: Various polynomials of degree 1 are

$(0 \text{ or } 1) \times x^1 + (0 \text{ or } 1) \times x^0$

i.e., $(0 \times x^1 \text{ or } 1 \times x^1) + (0 \times x^0 \text{ or } 1 \times x^0)$

i.e., $(0 \text{ or } x) + (0 \text{ or } 1)$

i.e., $(0 + 0) \text{ or } (0 + 1) \text{ or } (x + 0) \text{ or } (x + 1)$

i.e., $0 \text{ or } 1 \text{ or } x \text{ or } (x + 1)$

Disregarding the trivial choices 0 and 1, the Degree 1 polynomials are

x and $x + 1$

Degree 2: Note that just as we disregard the use of 1 as a factor in arithmetic calculations, so too we disregard the trivial polynomial 1 as a polynomial factor. Obviously, neither of the two polynomials of degree 2 (viz., x and $x + 1$) can be factored into two lower order, non-trivial, polynomials; so, by definition, the only two possible prime polynomials of degree 2 are:

x and $x + 1$

Degree 3: First, let us consider *all* polynomials of degree 3:

$$(0 \text{ or } 1) \times x^2 + (0 \text{ or } 1) \times x^1 + (0 \text{ or } 1) \times x^0$$

Omitting trivial polynomials, the polynomials of degree 3 are

$$x^2 = x \times x$$

$$x^2 + 1 = (x+1) \times (x+1)$$

$$x^2 + x = x \times (x+1)$$

$$x^2 + x + 1$$

Note that we have used the result

$$(x+1) \times (x+1) = x^2 + x + x + 1 = x^2 + 1$$

since $2x \to 0$ in modulo (mod) 2 (where $2 \to 0$).

From above, since the first three polynomials (i.e., x^2; $x^2 + 1$; $x^2 + x$) are expressible as two monic polynomials of degree 1 (i.e., x and $x + 1$), they are not prime polynomials.

To determine whether the fourth polynomial in the above set (viz., $x^2 + x + 1$) is a prime polynomial, we divide it by the two polynomials of lower order, viz., x and $x + 1$. If the resulting remainder is 0, then the polynomial (of degree 2) is completely divisible by the factor in question; otherwise, if the remainder is 1, then it is *not* completely divisible by the lower order polynomial.

Dividing $x^2 + x + 1$ by x,

$$
\begin{array}{r}
x\,)\ x^2 + x + 1\ (x + 1 \\
\underline{x^2} \\
x \\
\underline{x} \\
+1
\end{array}
$$

$(remainder = +1)$

Since the remainder in the above division is 1 (and not 0), x is *not* a factor of $x^2 + x + 1$.

Similarly, dividing $(x^2 + x + 1)$ by $(x + 1)$,

$$
\begin{array}{r}
x + 1)\ x^2 + x + 1\ (x + 0 \\
\underline{x^2 + x} \\
+\ 1
\end{array}
$$

$(remainder = +1)$

Thus, $(x + 1)$ is *not* a factor of $x^2 + x + 1$ either, since the remainder is 1 (and not 0).

It is to be concluded that $x^2 + x + 1$ is *not* divisible by either of the two lower order polynomials (viz., x and $x + 1$). Therefore the only prime polynomial of degree 2 is $(x^2 + x + 1)$.

Degree 3: Considering all the polynomials of degree 3 and trying to obtain their factor polynomials (of degree 2), we find that, in mod (2) arithmetic operation,

$$x^3 = x \times x \times x$$

$$x^3 + 1 = (x+1)(x^2 + x + 1)$$

$$x^3 + x = x \times (x^2 + 1)$$

$$x^3 + x + 1$$

$$x^3 + x^2 = x^2 \times (x+1)$$

$$x^3 + x^2 + 1$$

$$x^3 + x^2 + x = x(x^2 + x + 1)$$

$$x^3 + x^2 + x + 1 = (x^2 + 1) \times (x+1)$$

Disregarding the polynomials which can be expressed as a product of two (or more) lower degree polynomials, we observe that the only prime polynomials of degree 3 are:

$$x^3 + x + 1$$

and

$$x^3 + x^2 + 1$$

The above results are useful toward developing algebraic theory-based higher error-correcting codes with the use of prime polynomials.

6.11. Primitive Roots

Noting that if:

$$x^n = 1 = e^{i2\pi k}, (i = \sqrt{-1}) \qquad (6.7a)$$

then

$$k = 0, 1, 2, 3, \ldots n - 1 \qquad \text{(any integer} < n)$$

Note that if $k = n$ or $k > n$, then writing $k = n + l$, $l = 0, 1, 2, \ldots$, we get

$$e^{i2\pi(n+l)} = e^{i2\pi n} e^{i2\pi l}$$

$$= [\cos(2\pi n) + (i \sin 2\pi n)] \times [\cos(2\pi l) + i(\sin 2\pi l)]$$

$$= 1 \times 1 = 1$$

since, for integer n and l, $\cos(2\pi n) = \cos(2\pi l) = 1$ and $\sin(2\pi n) = \sin(2\pi l) = 0$.

Since we disregard the trivial factor 1, it suffices to consider integers $k \leq (n - 1)$.

The solutions or roots for x in equation (6.7a) is given by

$$x = e^{i2\pi k/n}, \quad k = 0, 1, 2, \ldots, n-1; n \neq 0 \quad (6.7b)$$

The successive powers of the above roots can be written as

$$(x)^m = e^{i2\pi km/n} \quad (6.7c)$$

If the value of the above expression for any two distinct powers (say m_1 and m_2, $m_1 \neq m_2$) are equal, i.e., if

$$e^{i2\pi km_1/n} = e^{i2\pi km_2/n}; \quad m_1, m_2 = 1, 2, 3, \ldots (m_1 \neq m_2)$$

or

$$e^{i2\pi k(m_1-m_2)/n} = 1$$

we must have

$$\frac{k(m_1-m_2)}{n} (\rightarrow \text{an integer}) = 0, 1, 2, \ldots$$

Since k is a prime (not divisible) to n, the above can hold only under the condition $m_1 = m_2$.

Therefore, the successive powers of the root (6.7a) generate all the roots of unity; and hence this root is called a *primitive root*.

6.12. Matrix Representation

A code of $n = 2^m - 1$ digits can be considered in terms of a matrix $M = (m, n)$, each column of which represent the n binary representation (abbreviated as $BR_i, i = 1, 2, \ldots n - 1$); as shown below:

```
              Col. 1   Col. 2   Col. 3 ...   Col. n
              BR₁      BR₂      BR₃          BRₙ
Row 1        |[ 0      0        0 ...        1 ]
Row 2        [|| 0     0        0 ...        1 ]
Row 3        [  0      0        0 ...        1 ]
.            [                                . ]    = M(m,n)
.            [                                . ]
.            [                                . ]
Row (m-2)    [  0      0        0 ...        1 ]
Row (m-1)    [  0      1        1 ...        1 ]    |
Row m        [  1      0        1 ...        1 ]
```

For instance, for $n = 7 = 2^3 - 1$, we have the matrix $M(3, 7)$

$$M(3,7) = \begin{bmatrix} 0 & 0 & 0 & 1 & 1 & 1 & 1 \\ 0 & 1 & 1 & 0 & 0 & 1 & 1 \\ 1 & 0 & 1 & 0 & 1 & 0 & 1 \end{bmatrix}$$

Such bit matrices could be used for coding purposes; and any interchange of the columns do not basically change the code. The particular column that matches the *syndrome* (see Sections 6.8 and 6.13) of code suitable for detecting a single bit error then reveals the error, which can then be corrected. Only the position and pattern of the matching column is important for such 1-bit error detection and correction.

Example[13]

Let us consider the prime polynomial (of degree 3):

$$x^3 + x + 1$$

Assuming that 'd' is a root of this polynomial, we have

$$d^3 + d + 1 = 0$$

i.e.,

$$d^3 = -(d + 1) = +(d + 1)$$

The last equality is written using the result

$$-(d + 1) = -2(d + 1) + (d + 1)$$
$$= 0 + (d + 1), \text{ (for mod 2)}$$
$$= d + 1$$

Using the relation (for mod 2 arithmetic)

$$\alpha^3 = (\alpha + 1)$$

powers of α equal to or higher than 3 can be reduced to sums of lower power representations of α, using matrix algebra (mod 2). Algebraically, we have expressions for $1 = \alpha^0$, α^2, etc., and converting these in matrix forms, we can write:

$$1 \to 0 \times \alpha^2 + 0 \times \alpha^1 + 1 \times \alpha^0 \to \begin{pmatrix} 0 \\ 0 \\ 1 \end{pmatrix}$$

Similarly,

$$\alpha \to 0 \times \alpha^2 + 1 \times \alpha^1 + 0 \times \alpha^0 \to \begin{pmatrix} 0 \\ 1 \\ 0 \end{pmatrix}$$

$$\alpha^2 = 1 \times \alpha^2 + 0 \times \alpha^1 + 0 \times \alpha^0 \to \begin{pmatrix} 1 \\ 0 \\ 0 \end{pmatrix}$$

Combining the foregoing results (after any factorization as necessary)

$$\alpha^3 = d + 1$$

$$\alpha^4 = \alpha^3 \times \alpha = (d+1) \times d$$

$$= \alpha^2 + \alpha$$

$$\alpha^5 = \alpha^3 \times \alpha^2 = (d+1) \times \alpha^2$$

$$= \alpha^3 + \alpha^2$$

$$\alpha^6 = \alpha^3 \times \alpha^3 = (d+1) \times (d+1) = \alpha^2 + 1$$

since $2\alpha \to 0$ (for mod 2),

i.e.,

$$\alpha^6 = \alpha^2 + 1$$

$$\alpha^7 = \alpha^6 \times \alpha = (\alpha^2 + 1) \times \alpha$$

$$= \alpha^3 + \alpha$$

for mod 2 $(1 + 1 = 2 = 0)$

which is the same as the matrix representation of 1.

Let us collect the above (3 x 1) matrices in the form of a single (3 x 7) matrix, M^1.

By interchanging columns as shown by the double arrows (note that a set of two interchanges is required for the 7th column as indicated by two layered curved double arrows), the matrix M^1 can be transformed into the original 3 x 7 matrix M. Further, by reversing the order of all the 7 columns of M, we can rearrange M^1 into the matrix M^{11}.

At the receive end, we compute the syndrome and compare it with the successive powers of α. When a match between the syndrome and a column is obtained, the decoding function can be performed.

6.13. Syndrome

Consider the bits to be transmitted in a linear fashion (i.e., in one dimension), with parity check bit appended to the set. Alternatively, a rectangular (2-dimensional coding), or a cubic (3-dimensional) coding can also be perceived. In fact, even higher-dimensional coding can also be imagined. In any particular arrangement, the ratio:

$$R = \frac{Total\ number\ of\ bits\ (n)}{Number\ of\ information\ bits\ (k)}$$

is called the redundancy. If the number of information bits is k and the total number of bits is n (>k) (i.e., number of parity check bits is $n - k$), then we have their ratio

$$R = \frac{n}{k} = \frac{k+(n-k)}{k} = 1 + \frac{n-k}{k} \qquad (6.9a)$$

More generally, if the total number of bits is 2^n, of which $(n + 1)$ bits are parity check bits, then, if the parity bits are suitably arranged, they would give a number called *syndrome*. The syndrome can be obtained by writing a 1 for each failure of the parity bits (parity *not* satisfied), and a 0 for each successful parity (when parity *is* satisfied).

The resulting $(n + 1)$ bits by writing 1s and 0s as prescribed above form a $(n + 1)$-bit number, and this number, called the *syndrome* as mentioned above can specify 2^{n+1} objects. These include the n locations where a single error occurred, plus the fact that no error occurred.

The syndrome, in fact, contains more than enough states to indicate when the entire message is correct, plus the position of the error if one occurs. This implies a loss of efficiency. Nevertheless, the syndrome provides a new approach to design of an error-correcting code.

6.14. Coding Efficiency

Different types of measures can be employed to assess the efficiency of the coding process. Such measures include redundancy, coding gain, and throughout.

Redundancy

If the number of the information bits is k and the number of coded bits is n, this is referred to (k, n) coding and the ratio of n to k is termed the *redundancy* (R):

$$R = \frac{n}{k} = \frac{k+(n-k)}{k} = 1 + \frac{n-k}{k} \qquad (6.9b)$$

The radio $(n - k)/k$ is termed as the *extra redundancy*.

Obviously, the redundancy directly impacts the bandwidth requirement, since the coding bits imply use of additional bandwidth compared to the bandwidth required to transmit the un-coded message.

Coding Gain

Use of coding expectedly reduces or eliminates the bit error induced during the transmission across the channel. This advantage is quantified in terms of improvement in the energy-per-bit to noise-density ratio, by defining the parameter *coding gain* (G_c) as the ratio

$$G_c = (E_b/N_o)_u / (E_b/N_o)_c \qquad (6.9c)$$

where $(E_b/N_o)_u$ is the energy-per-bit-to-the noise-power-density ratio for transmission of the uncoded message word, and $(E_b/N_o)_c$ is the analogous ratio for the coded word, to give the same amount of bit error rate at a receiver, assuming the rate of information transmission remains the same in both cases. Since the coded message has a larger number of bits compared to the uncoded message, a constant rate for information transmission clearly implies that the bit rate of the coded signal is higher than that of the uncoded signal. For the transmission of the uncoded and coded signals, if the bit rates are r and r', respectively, then for a (n, k) coding, we have

$$r = \frac{k}{t_o}; \text{ and } r' = \frac{n}{t_o} \qquad (6.10a)$$

where the time duration t_o, must remain the same in the two cases. Hence, (since $n > k$), the difference in the bit rates, Δr, is

$$\Delta r = r' - r = \frac{n-k}{t_o} \qquad (6.10b)$$

But, by definition of the redundancy, R (= n/k),

$$R - 1 = \frac{n-k}{k}$$

i.e.,

$$n - k = k(R - 1)$$

Hence,

$$\Delta r = \frac{k(R-1)}{t_o} = r(R-1) \qquad (6.10c)$$

The above equation provides a relation between the increase in the bit rate and the original (uncoded) bit rate and the redundancy due to coding.

For a Gaussian channel, the coding gain can be approximated[13] as follows:

$$G_c = 10 \log_{10}\left(\frac{d}{R}\right) \tag{6.11}$$

where $R = \frac{n}{k}$ is the *redundancy* (note: the inverse ratio, $\frac{1}{R} = \frac{k}{n}$, is the *code rate*) for this (k, n) coding, and d is the minimum code distance for the block or convolution code. Typically, coding gains in the range of 4-7 dB is achieved in practical cases.

Note that if the code distance (d) is increased, the coding gain (G_c) increases, implying a decrease in the error rate, in consistence with Shanon's theorem *("The error rate can be made arbitrarily small by increasing the code length")*. The error rate decreases by increasing the block size (or constraint length). However, increasing the code length implies a greater level of complexity for the decoding process (i.e., the decoder-complexity is increased).

6.15. Coding in Satellite Communications[14-16]

Satellite communications are power-limited, particularly for small earth stations. Mobile terminals and Very Small Aperture Terminals (VSATs) have to operate in severely power- limited conditions, and normally would suffer severe signal degradation, which can be mitigated only with the use of coding. The propagation medium, particularly in the case of mobile communications, may be quite unfavorable or even hostile, causing excessive degradation; and, again, coding is the only recourse for appropriate level of improvement in the signal quality at the receive end.

In the case of computer communications and data transmission, virtually an error-free transmission is essentially required. Hence, for satellite links for such stringent cases and applications, error-free transmission can be ensured only through coding. In general, coding can potentially improve the overall channel capacity, as well.

Due to long transmission time (delay) involved in satellite channels, particularly in the case of geosynchronous satellites, ARQ type of coding is impractical, and the FEC coding is typically employed in satellite communication applications. Under adverse propagation conditions such as rain, fog, lightening, urban atmospheric pollution, etc., occurrence of burst- errors is naturally expected, and the choice of the type of FEC error correcting code must take into account such situations.

As coding is applicable only for digital (baseband) signals, it has become a general trend that satellite communications for various applications (voice, television, and of course, data transmission) is performed in digital mode, with distinct advantage over analog transmission. Thus the use of analog (baseband) signal transmission for voice and television via satellite links has been rapidly diminishing in recent years; digital communications being implemented for these applications in rapidly increasing rate.

Use of digital signal also provides a number of advantages including signal quality and channel capacity enhancements, and flexibility of mixing and manipulating the data to provide special applications ("APPs"). A well familiar case is digital TV news channels with multiple layers of text materials, some in the "streaming mode." In recent years, the variety and number of APPs offered to mobile and cellular telephone, including Internet capability, smart and intelligent hand-held receivers, introduced in the industry for large-scale public use, is all too familiar to everyone.

References:

[11] C.E. Shanon, "A Mathematical Theory of Communications," Bell System Technical Journal (BSTJ), vol. 27, p. 379-623.
[12] R.W. Hamming, Coding and Information Theory, Prentice Hall, Englewood Cliff, NJ, 1980.

[13] P.G. Farrel and A.P. Clark, Modulation and Coding, International Journal of Satellite Communications, Vol. 2, 1984, pp. 287-304.
[14] J. A. Heller and I.M. Jacobs, Viterbi Decoding for Satellite and Space Communications, IEEE Trans. Communication Technology, COM-19, October 1971, pp. 835-848.
[15] W.W. Wu, Applications of Error-Coding Techniques to Satellite Communications, Comsat Technical Review, Vol. 1, Fall 1971, pp. 183-219.
[16] Richard W. Hamming, Coding and Information Theory, Prentice Hall, Englewood Cliff, NJ, p. 206, 1980.

7. CODING – II

7.1. Types of Coding

Mainly two classes of codes are used in satellite communications: *block codes* and *convolution codes*. The basic principles underlying these two coding methods are briefly described below.

7.2. Block Codes

In a (n, k) coding (i.e., k information bits and n coded bits; with $n - k = r$ redundant bits), the n bits may be treated as a block, yielding what is referred to as *block codes*.

In any block of code word, a block code can

(a) <u>detect</u> up to e_D bits in error, where

$$e_D < (d_H - 1) \tag{7.1a}$$

where d_H is the Hamming distance in a set of such code words. For a minimum value e_{Dm} $(= 1)$ of e_D (i.e., for the block code to be able to detect at least one bit error), we must have

$$d_H = e_{Dm} + 1 = 1 + 1 = 2 \tag{7.1b}$$

i.e., the Hamming distance must be 2.

(b) <u>detect and correct</u> e_{DC} bits in error, we must have

$$e_{DC} \leq (d_H - 1)/2 \qquad (7.2a)$$

or,

$$d_H \geq [2e_{DC}] + 1 \qquad (7.2b)$$

where the brackets [] on the right means that if e_{nc} as given by equation (7.2a) yields a non-integer value then e_{DC} value used in equation (7.2b) is taken to be the next lower integer. Thus, if the block code is to detect 5 bit errors and correct 2 bit errors, then the Hamming distance must be at least 5.

More generally, if the block is to detect e_D errors and correct e_C errors ($e_C \leq e_D$), then the Hamming distance is given as

$$d_H = e_D + e_C + 1 \qquad (7.2c)$$

The implementation of block codes usually makes use of the linear algebraic theory based on linear transformation of the code word into a polynomial and into matrix representation on the basis of mod 2 arithmetic as discussed in Sections (6.10 – 6.13). An alternative to this makes use of *orthogonal signals*. For longer code words, the decoding process for orthogonal signals becomes increasingly more complex; and comparatively the simpler alternative (algebraic theoretic coding with linear matrix algebra) is therefore commonly used.

In order to formulate the coding-decoding process mathematically (in terms of matrix algebra), suppose a k-bit un-coded code word which has bits (0 or 1) represented by ($b_1, b_2, b_3, ... b_k$). It is then convenient for block coding to represent this sequence as a (1-row x k-column) matrix (or a vector) B:

$$B = [b_1, b_2, b_3, ... b_k] \qquad (7.3a)$$

We then assume that the code word C is generated by using a suitable *generation matrix* M by forming a matrix multiplication, i.e.,

$$C = BM \qquad (7.3b)$$

The process of code system design then aims to determine the matrix M consistent with the general objective of minimizing the redundancy and obtaining the specified level of error probability for a given channel noise environment. Two approaches are possible, leading to:

(i) <u>systemic codes</u> – when the k information bits are preserved in their original form, so that an estimate of these information bits can be easily made by observation; and
(ii) <u>non-systemic codes</u> – when the information bits are not explicit and hence their estimation is not feasible by observation.

In general, all combinations of the k information bits can be directly generated and the decoder uses a *reference (look-up) table* of such 2^k possible vectors known to the coder and decoder both. In the case of systematic codes, assuming r redundant bits are used, the 2^r possible rows for r redundant bits suffices to provide the reference (look-up) table, simplifying the decoder design by reducing its complexity and increasing efficiency considerably. The generator matrix can be written as:

$$M = IK \qquad (7.3c)$$

where I is a $(k \times k)$ identity matrix (with k bits in each row and in each column):

$$I = \begin{matrix} [1 & 0 & 0 \ldots 0] \\ [0 & 1 & 0 \ldots 0] \\ [0 & 0 & 1 \ldots 0] \\ & \cdot & \\ [0 & 0 & 0 \ldots 1] \end{matrix}$$

$$(7.3d)$$

and K is a kxr matrix which when suitably designed, yields M (Equation 7.3c), and therefore, C (Equation 7.3b).

The decoding process is based on the matrix relationship:

$$PC^T = 0 \qquad (7.4a)$$

where C^T is the transpose matrix of C, and P is a known matrix, used by the decoder.

In general, the decoder first obtains the syndrome S of the received code word, by using corrupted version C^1 of C, by using the relationship:

$$S = PC^1 \qquad (7.4b)$$

Generally, as C^1 contains bit-errors as compared to C due to corruption in the channel. the syndrome S would be non-zero. If with the corrupted version C^1, the following relation is satisfied with respect to the transpose matrix Q^T

$$S = PQ^T \neq 0 \qquad (7.4c)$$

where Q is called the *error matrix* and is known to the decoder; then the non-zero element(s) of S obtained from equation (7.4c) represents a bit error which can then be corrected. The decoded word is finally obtained from the relation

$$C = C^1 \oplus Q \qquad (7.4d)$$

The bounds of error detection and correction are given by the Hamming distance. The vector Q can be estimated by solving for it, starting with one bit error, up to the maximum limit based on the Hamming distance. Error probabilities are adjusted to acceptable levels by adjusting the power level (i.e., by proper link design).

Efficiency of Block Coding

In a (n, k) block coding, there are 2^k possible *message words* (all possible combinations of k information bits, each bit-position permitting two possible choices: 1 or 0); while the total number of possible received words formed by n bits is obviously 2^n.

Only one out of 2^k message words is valid code word. On the other hand, bit erorrs can cause the received code words to be interpreted as one of the spurious code words. The number of such false detections is therefore $(2^k - 1)$. Such a false reception may be generally very low in a good decoder design. Still it is useful to define a probability of bit errors as the ratio of the number of possible detections $(2^k - 1)$ to the total number of received words (2^n). Actually, due to low occurrence of false detections, this ratio represents the upper bound of the error probability, P_e, i.e.,

$$P_e \leq (2^k - 1)/2^n \tag{7.5a}$$

The error probability (P_e) defined above obviously decreases by choosing coding (n, k) with lower value of the information bits (k), and a higher value of the total number of coded bits (n). Values of the error probability P_e of the order of 10^{-10} or lower can be considered very effective block coding. The error probability of lower values can be obtained by increasing the energy-per-bit-to-noise density ratio (E_b/N_o). Depending on the specific choice of modulation and coding techniques, for a range of E_b/N_o of 5-10 dB, the P_e value in the range $10^{-3} - 10^{-5}$ may be achieved.[17]

Throughput

In a block code using the automatic receive request (ARQ), if the number of information bits received per unit time is: k_w with ARQ, and k_o without ARQ, then the ratio

$$R_Q = k_w/k_o \qquad (7.5b)$$

is referred to as the *throughput* of the system. Table 7.1 summarizes the common features of the block coding with FEC.

Table 7.1
Features of Major Block Coding Schemes with FEC

Code Name	Block Length in bits (n)	No. of information bits (k)	No. of Parity (coding) bits ($n-k$)	Hamming (minimum) distance (d)	Remark
Hamming	$2^m - 1$		m	3	
BCH	$2^m - 1$ $m = 3, 4, ...$	$\geq n - mt$		$\geq 2t + 1$	Cyclic
R-S	$2^m - 1$ (s) $= m(2^m - 1)$		$2t(s)$ $= m \times 2t$	$2t + 1(s)$	m bits/symbol(s)
Maximum-Length (simplex)	$2^m - 1$	m		$2^m - 1$	Spread-Spectrum
Quadratic Residue	p $= 8m \pm 1$	$\dfrac{p+1}{2}$		$> \sqrt{9}$	Cyclic ($p \to$ a prime number)
Golay[18]	23 (+1 parity – 24)	12	*	*	3-error-correction

7.3. Cyclic Codes [19]

If one connects the start and the end stages of a shift register, then the codeword could be generated by shifting the bits in a circular or cyclic configuration. The bit-shifting can occur in this configuration either in a clockwise or in an anti-clockwise fashion. The coding-decoding process in this cyclic manner are classified as *cyclic codes*, and the process involved in this class of coding helps to simplify the decoding process (reducing the decoder complexity).

A number of well-known cyclic codes have been 'invented' including:

(i) Hamming code
(ii) Reed-Solomon Code (RS Code)
(iii) Bose, Chaudhari and Hocquenghem (BCH)
(iv) Golay Code

The RS code and the BCH codes are often used in satellite communications systems. In general, cyclic codes form an important class of codes for application in satellite communications.

In a cyclic code, the message, $B(x)$, compromising the k information bits is mapped onto a polynomial of degree $k-1$, as follows

$$B(x) = B_o + B_1 x + B_2 x^2 + \cdots + B_{k-1} x^{k-1} \qquad (7.6a)$$

where B_i is the coefficient of x^i ($i = 0, 1, 2, \ldots k-1$), and all the additions on the right of equation (7.6a) are mod-2 additions.

The code word $C(n)$, comprising $n = k + r$ bits in the (n, k) coding can also be respresented as a polynomial

$$C(n) = C_o + C_1 x + C_2 x^2 + \ldots + C_{n-1} x^{n-1} \qquad (7.6b)$$

where the additions are again mod-2 additions. Introducing the generating function $G(x)$, such that

$$C(n) = G(x)\, B(x) \qquad (7.7a)$$

one can write for the generating function:

$$G(x) = 1 + G_1 x + G_2 x^2 + \cdots + G_{r-1} x^{r-1} + x^r \qquad (7.8b)$$

where Gi is the coefficient of x^i, ($i = 0, 1, 2, \ldots, r$), but $G_r = 1$, and also assuming again all the additions on the right of equation (7.7b) to be mod-2 additions. The coefficients of $G(x)$ are obtained by factoring,

$$F(x) = 1 + x^n \qquad (7.8a)$$

assuming again the addition on the right of Equation (7.8a) to be mod-2.

The factor Equation (7.8a) having (n, k) degree of freedom [which is alos the degree of freedom of the generating polynomial, since $B(k)$ has a degree of freedom $k - 1$ and $C(n)$ has a degree of freedom $n - 1$; so from equation (7.7a), $G(x)$ must have a degree of freedom equal to

$$(n - 1) - (k - 1) = n - k$$

G(x) is therefore taken to be the code-generating polynomial.

Cyclic Hamming Code

Cyclic Hamming Codes are based on the Hamming distance (e_D) as the basic parameter.

From the relation (Equation 7.2c)

$$d_H = e_D + e_c + 1 \qquad (7.8b)$$

For a (n, k) code word with $n - k = r$ bits being redundant bits for coding, if $d_H = 3$, then

$$e_D + e_c = d_H - 1 = 3 - 1 = 2 \qquad (7.8c)$$

Also, for $r = 3$, a solution is $n = 7$, and $k = 4$, a cyclic hamming code can detect 1 error, and correct 1 error.

7.4. Reed Solomon Code (RS Code)

RS code operates over symbols rather than over bits. Thus, in case of RS coding $n - k$ implies that the information bits are contained in k symbols to yield code word of n symbols, with the $r = n - k$ symbols being used as coding symbols or parity symbols, and k/n being the code rate. If one or more bit errors occur in a symbol, the symbol is said to be in error.

Assuming that the number of bits per symbol is m, the number of symbols per code word is given as

$$n = 2^m - 1 \qquad (7.9)$$

An RS code can correct errors in $r/2$ symbols.

Example:

Suppose we have 8 bits per symbol. i.e., $m = 8$. Then from equation (7.9), the code word has $2^8 - 1 = 255$. If we wish to correct 8 symbol errors, then $r/2 = 8$, or $r = 16$. Consequently, k= n−r =255−16 =239, and one can useand RS coding with (n,k) = (255,239).

(a-a1) rows are occupied by code bits, then clearly,

$$k = a_1 \times b \qquad (7.10\text{b})$$

and

$$r = (a - a_1) \times b \qquad (7.10\text{c})$$

In the presence of burst errors due to impulse noise, for example, RS coding offers special advantage. If a code can correct β errors, then the maximum length of noise burst that can be corrected is given by

$$\beta m \le \beta \text{ bits} \qquad (7.11\text{a})$$

Under interleaving, the number of consecutive bits that can be corrected increases by a factor m, the number of bits per symbol. Thus the number of burst errors that can be corrected can be written as

$$\beta m = \beta \qquad (7.11\text{b})$$

RS coding is also applicable to the case of multilevel modulation. The efficiency of RS coding decreases for random errors, however. If burst errors coexist with random errors, the RS coding can be used in conjunction with a suitable random error correction scheme. Such a hybrid error correction scheme is referred to as *concatenation*. In such an arrangement, the RS coding applied column-wise corrects the burst errors, while the supplementary error correction code is applied row-wise to correct random bit errors. Generally, relatively large periods of noise-free (i.e., error-free) signal must separate consecutive bursts for an efficient working of RS codes.

7.5. Turbo Code

For optimum performance, an arrangement referred to as turbo coding[20] is applied. In this arrangement, two parallel RS codes are applied. In such a case, RS coding is applied at the outer side, in concatenation with a convolution code applied in the inner side.

Simple row versus column interleaving scheme characteristic of block codes allows simple decoding scheme. The decoder in the TCP decoder is smaller in size than in the case of CTC decoder, and less complex to implement. At the same time, the TCP decoder is more energy-efficient and lower in cost.

As in the case of 3G implementation, turbo codes are capable of supporting high data rate requirements of the 4G systems, as well.

7.6. Low Density Parity Check (LDPC) Codes[22]

Improvement in coding scheme, even beyond the turbo codes, is possible by using the LDPC codes, which in combination with the FEC coding, yield performance closer to the Shanon limit.

The LDPC code is a linear error correcting code featuring a parity check matrix with a small number of non-zero elements in each row and column. In a most commonly used form, these non-zero elements are simply "1's." To describe how the LDPC coding works, it is useful to first define a partitioned generator matrix, G; a parity check matrix, H; and a transformation matrix, P; as follows:

$$H = [P^T : I] \qquad (7.12a)$$

and

$$G = [I : P] \qquad (7.12b)$$

where P^T is the transpose matrix of P and I is the identity matrix. As an example, assuming

$$P = \begin{matrix}[1 & 1 & 1] \\ [1 & 1 & 0] \\ [1 & 0 & 1] \\ [0 & 1 & 1]\end{matrix}$$

$$(7.13a)$$

we can write

$$G = \begin{bmatrix} 1 & 0 & 0 & 0 & : & 1 & 1 & 1 \\ 0 & 1 & 0 & 0 & : & 1 & 1 & 0 \\ 0 & 0 & 1 & 0 & : & 1 & 0 & 1 \\ 0 & 0 & 0 & 1 & : & 0 & 1 & 1 \end{bmatrix}$$

(7.13b)

and

$$H = \begin{bmatrix} 1 & 1 & 1 & 0 & : & 1 & 0 & 0 \\ 1 & 1 & 0 & 1 & : & 0 & 1 & 0 \\ 1 & 0 & 1 & 1 & : & 0 & 0 & 1 \end{bmatrix}$$

(7.13c)

Then the LDPC codes, C, for a data block X are generated as follows:

$$C = XG = [X:XP] \qquad (7.14a)$$

For all valid codes (G), the following matrix identity is satisfied:

$$C_y H^T = 0 \qquad (7.14b)$$

where C_y is the mod-2 sum of each row thus equals zero (0).

It should be noted that the matrix Equation (7.13b) corresponds to the so-called *Gauss elimination method* commonly used in linear algebra for solving a set of simultaneous linear equations. Thus the LDPC coding technique uses the Gauss elimination method for a direct way to obtain the LDPC coding and to perform the encoding process.

In general, the H matrix is a $[(n-k), n]$ matrix, where n is the total number of bits, out of which the number of message bits is k, so $(n-k)$ is the number of parity bits in this (k/n) code rate case.

7.7. Convolution Coding

Block coding involves the coding based on a block structure of input data, with code words created block-by-block, without interaction or mingling of the bits of one block with the bits of another block. In contrast, in convolution coding, coding of the code word involves bits of the one or more preceding code word as well, according to a pre-arranged algorithm. The selected bits of the previous code words consistent with the chosen algorithm are kept in the memory to be used in conjunction with the bits of the current code word and the set of bits are used in the shift register to implement the coding process.

The shift register thus convolves the information bits with the bits of the previous code word. In this convolution coding, the impulse response of the shift register – defined as the encoder response when a characteristic bit sequence 1000...0 (i.e., a single 1 followed by a set of 0s) – plays a special role in the coding process.

One or more adders, performing the linear exclusive – OR addition at various stages within or outside of the shift register also, obviously, play a basic and critical role in the coding process.

The number of information bits that affect the encoder output is referred to as the *constraint length* of the code. By virtue of the pattern of convolution coding, the constraint length is equal to the number of stages of the shift register, which is equal to the number of information bits that influence the encoder output. Clearly, unlike block coding where a block of bits being input or output of the shift register forming the code word is defined unambiguously, convolution coding is based on a continuous flow of bits coded according to the specific algorithm involving feedback and feed-forward process.

However, it is useful to determine the *minimum distance*, d_{min}, analogous to the hamming distance in the case of block code. For this process, the constraint length of the code is taken as the block size. Then, different valid code words can be compared in a manner similar to the code words in block coding. In this manner of treatment, it is possible to alter certain selected bits of one code word to reproduce another code word. One can therefore also determine the minimum number of bits by which one code word differs from another. This minimum number is then defined as the minimum distance, d_{min}, of the convolution coding. In contrast, it is also conventional to define a *free distance*, d_f, for convolution coding when a semi-infinite (very large bit sequence $\gg$ constraint length) sequence of code is considered.

An increase in the constraint length increases the error-correction capability; but an increase in the constraint length (i.e., an increase in the shift register stages) also increases the decoding complexity. This trade-off implies a suitable compromise in the error-correction capability versus the decoder complexity. An optimum design guideline under this type of compromise is usually performed by means of computer simulation.

Two types of approaches in the decoding process are referred to as '*hard decision*' and '*soft decision*', respectively. In the hard decision case, the code word is formed by taking decision on a bit-by-bit basis; while in the soft decision case, the demodulated analog signal is first correlated with locally generated binary sequences, and the binary sequence yielding the highest correlation is taken as the detected binary stream. Cleary the soft decision approach implies a greater degree of complexity in the design and operation of the decoder. However, this soft decision approach provides improved (2 dB) efficiency and accuracy in comparison to hard decision.

Convolution coding is a preferred choice since in satellite communications the large propagation delay renders the block coding impractical for satellite network. Convolution coding applied to satellite communications usually employs *Viterbi decoder*.

7.7. Viterbi Decoding[23-25]

Discarding the non-optimal paths (giving bit errors higher than the minimum value) greatly reduces the decoder complexity. This process (discarding non-optimal paths) is found to produce negligible degradation in the decoder performance. Thus Viterbi decoding proves to be the most cost effective choice for satellite communications.

Viterbi decoding is based on the concept of the *free distance*. The free distance is analogous to the minimum Hamming distance in the context of convolution code wit appropriate modification, since Hamming distance generally applies only to block codes. Thus the free distance is equal to the minimum number of bits that must be changed in order to form a valid code word in a semi-infinite sequence.

References:

[16] M. Richharia, Satelite Communications Systems, McGraw Hill, p. 183. 1999.

[17] Richard W. Hamming, Coding and Information Theory, Prentice Hall, 1980.

[18] C. Berrou and A. Glavieux, Near-optimum Error Correcting Coding and Decoding: Turbo-Codes, IEEE Trans. Communication, Vol. 44, No. 10, pp. 1261-1271. 1996.

[19] C. Berrou, A. Glavieux, and P. Thitimajshima, Near Shanon Limit Error Correcting Coding and Decoding: Turbo Codes, Proceedings of ICC '93, Geneva, May 1993, pp. 1064-1070.

[20] R.M. Tanner, A Recursive Approach to Low Complexity, IEEE Trans. Info., Theory, Sept. pp.533-547, 1981.

[21] M. Luby et al., Improved Low-Density Parity Check Codes Using Irregular Graphs, IEEE Trans. Info., Theory, Feb. 2001.

[22] T. Richardson and R. Urbanke, Effective Encoding of Low-Density Parity Check Codes, IEEE Trans. Info., Theory, Vol. 47, pp.638-656, 2001.

[23] Willian C.Y. Lee, Wireless and Cellular Telecommunications, p.744, 2006.

[24] A.J. Viterbi, Convolution Codes and Their Performance in Communication Systems, IEEE Trans. Communication Technology, COM-19, October 1971, pp. 751- 772.

[25] J.A. Heller and I.M. Jacobs, Viterbi Decoding for Satellite and Space Communication, IEEE Trans. Communication Technology, COM-19, October 1971, pp. 835-848.

8. FREQUENCY BAND ALLOCATION: INTER-SYSTEM INTERFERENCE (II)

Consideration of the Intersystem-Interference (II) is an important part of the system planning, design, and analysis. The presence of neighboring satellite(s), on one or both sides of the satellite under consideration, in the orbit, can cause II due to the so-called *off-axis* antenna gain of the antenna(s) of the interfering satellite(s). II will cause interference noise for the receive earth station(s) – that is the desired downlink. Similarly, the off-axis radiation pattern or gain of the transmitting earth stations of the neighboring satellite system(s), which are located within the coverage region of the intended satellite, will be received by the antenna(s) of the intended satellite and act as noise into the desired uplink.

In the planning stage of each satellite system, care is taken to minimize II from, ant to, other systems by reducing off-axis radiation of earth station and satellite antennas; but a total elimination of off-axis radiation may not be possible due to finite size antennas which inherently suffer from diffraction pattern and radiation leakage at the edges. Thus analysis and consideration of unavoidable uplink and downlink II usually becomes an essential part of system planning, unless other systems operating in the same frequency band are absent. With crowding of the geostationary orbit with multiple satellite systems (or for terrestrial mobile systems) operating in the same or overlapping frequency bands, consideration of II becomes increasingly more important. Any uplink and downlink II must be evaluated and factored into the calculation of the desired link performance.

The signal-to-interference noise power ration is denoted by (C/I). Thus, the uplink and downlink contributions (denoted here by the subscripts u and d), $(C/I)_u$ and $(C/I)_d$, respectively, need to be taken into account. If there are multiple sources of uplink and downlink II, each contribution has to be determined and the resultant uplink and downlink interference power ratios need to be evaluated using the usual formula, viz.

$$[(C/I)_u]^{-1} = \mathbf{S}[(C/I)_{uj}]^{-1} \qquad (8.1a)$$
$$\phantom{[(C/I)_u]^{-1} = }j$$

where the left hand side of Equation (8.1a) is the resultant value of the uplink power-to-II-noise ratio; the $(C/I)_{uj}$-term on the right is the contribution by the jth uplink II noise; and the summation is carried out over all sources (j=1,2,3, . . .) of uplink II. An analogous equation also holds for the downlink II sources and the resultant downlink II $(C/I)_d$.

Finally, the uplink and downlink carrier-to-II-noise ratios are computed by combining these values with the corresponding carrier-to-noise ratio, (C/N) computed without intersystem interference, again with analogous formulae:

$$[(C/N)_u]^{-1} = [(C/N)_u]_o^{-1} + [(C/I)_u]^{-1} \qquad (8.1b)$$

$$[(C/N)d]^{-1} = [(C/N)d]_o^{-1} + [(C/I)d]^{-1} \qquad (8.1c)$$

where the subscript o on the right denotes value without II. This leads to a more realistic and accurate value of the uplink and downlink carrier-to-noise ratios (the left sides of Equations 8.1b and 8.1c, respectively) in the presence of the II. Thus, an estimate of the link performance to be expected in actual practice can be made.

The specifications of the off-axis radiation (or gain) performance of all the satellite and earth station antennas involved, as a function of the off-axis angle, is provided with every antenna by the antenna manufacturer. It depends on the type of antenna, its size (diameter), smoothness of the antenna surface, location of the antenna feed assembly at or near the focal point of the antenna, and other similar factors. Large antennas have narrow beam and relatively lower value of the off-axis radiation. Commonly, antennas gains are of the (Sin^2x/x^2)-form (x being the off-axis angle); with one main lobe providing the desired radiation, while the second and higher order lobes contribute to intersystem interference, with diminishing magnitudes characteristic of diffraction patterns resulting from a finite-size opening.

In passing, it may be noted that similar diffraction patterns in the optical part of the spectrum can be easily observed on a small scale when the radiation is allowed to pass through a small slit. In the presence of two or more slits, interference fringes are formed. In fact, these effects (diffraction and interference) testify the wave-nature of radiation; whereas the photoelectric effect is explained by assuming a quantum or particle nature (photon) of radiation, as first done by Einstein who was awarded a Nobel Prize for it. The wave-particle duality, first propounded by DeBroglie, is more apparent in the quantum (optical and ultra-violet) domain. In the microwave domain (relevant for satellite and mobile telecommunications), the frequency is comparatively much smaller; i.e., the wavelengths are much larger, and the wave-nature manifestation predominates.

In the case of digital communications, obviously, the above type of intersystem interference analysis (Equations 8.1a,b,c) must be done in terms of the E_b/N_o (energy-per-bit-to-noise-power-density) ratio, rather than S/N or C/N ratios discussed above. The uplink and downlink values of this ratio attributed to the pertinent sources of II are to be evaluated to obtain the resultant link performance to ensure that the overall systems requirements specifications regarding network service quality are adequately met.

Often, the above type of analysis may lead to systems constraints – regarding the spacing of the neighboring satellite locations and/or the antenna radiation and gain pattern (specially the off-axis emission characteristics) of the satellite and earth station antennas. More specifically, small earth stations which have a wide beam and off-axis radiation may become of great concern to neighboring networks.

Consideration of both the desired and transverse polarizations must be made for this purpose. This implies suitable specifications for the cross-polarized emission limits. Thus a number of antenna systems performance guidelines are involved in addressing the intersystem interference. These considerations are part of the systems design process both in case of satellite telecommunications and terrestrial (mobile) telecommunications.

9. CELLULAR (MOBILE) COMMUNICATION SYSTEMS

9.1. Introduction

In early days of satellite telecommunications, the mobile satellite telecommunications primarily consisted of maritime mobile services (MMS), with somewhat different conventions for the European system (Global Satellite Maritime, or GSM) and American system (MS). The mobile system for individual personal mobile units came into being in the mid-1990s, and soon became one of the most popular methods of communications.

The new and globally ubiquitous mobile telecommunications systems used by individuals with pocket-size iPhones and other variations have evolved very rapidly in technology, design, protocol conventions, and capabilities or service applications (APPs). Various stages of this evolution is marked by the typical nomenclature BNG, where the middle character (N) is a number (N=2,3,4, . . .) and refers to the sequence of generation. Thus BNG is an acronym for "Beyond Nth Generation"). In this Chapter, we briefly review various important systems characteristics of successive generational mobile telecommunications systems. As in the previous Chapters, we confine ourselves to the basic conceptual and theoretical aspects only, without going into equipment design, implementation, operation, and other engineering aspects. The reader is referred to other suitable sources for such engineering details.

In the following Sections, we start our review with B2G ("Beyond 2G or Second Generation") systems, and proceed to B3G, B4G, etc.

9.2. The B2G Systems[26]

A number of wireless/mobile systems were developed *after* introducing TDMA and CDMA modulation schemes. These systems include GPRS, EDGE, HSCSD, iDEN, PHS, and RTTIX (IS-95B). These systems are collectively referred to as B2G systems. The GPRS system is briefly described below. The B2G system is now preempted by higher level systems in most networks.

General Packet Radio Service (GPRS)

The GPRS system is an improvement over the Global System for Mobile (GSM), and offers more efficient packet-based data services at higher data rates. One particular implementation – the WCDMA involving usage of GPRS within the UMTS Release 99 -- is an example. Packet-switching technology is used to allow sharing of the air-interface resources.

GPRS uses a 200-kHz channel divided into 8 time-slots (this is the same basic air interface as GSM), and four different channel coding schemes, CS-1, CS-2, CS-3, and CS-4, each with a different data rate. The CS-2 with a data transmission rate of 13.4 kbps and a coding scheme for packet data transfer is the most common. The actual data rate is approximately 20-30 percent less than the transmission rate. The given time slot is called packet data channel (PDCH); while PCCCH and PDCCH defined below represent logical channels, and improvements over that GSM, GPRS and GSM having similar basic structure.

The PCCCH comprises:

(1) Packet Random Access Channel (PRACH) to initiate a transfer of packet signaling or data in the uplink;

(2) A Packet Paging Channel (PPCH) to page an MS prior to a downlink packet transfer;
(3) A Packet Access Grant Channel (PAGCH) to be used by the network to assign resources to the MS, prior to the downlink packet transfer;
(4) A Point-to-Multipoint Multicast (PTMM) channel called the Packet Notification Channel (PNCH) to send notification to a group of MS.

A Radio Block (RB) consisting of 4 TDMA frames is used. The GRPS air interface frame structure is shown in Figure 9.1.

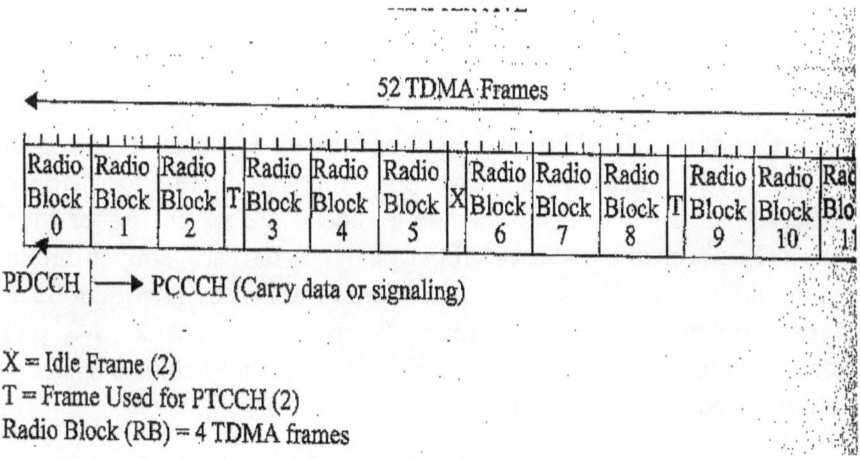

Radio Block 0	Radio Block 1	Radio Block 2	T	Radio Block 3	Radio Block 4	Radio Block 5	X	Radio Block 6	Radio Block 7	Radio Block 8	T	Radio Block 9	Radio Block 10	Radio Block 11

← 52 TDMA Frames

PDCCH ⟶ PCCCH (Carry data or signaling)

X = Idle Frame (2)
T = Frame Used for PTCCH (2)
Radio Block (RB) = 4 TDMA frames

Figure 9.1. GRPS Air Interface Frame Structure—only10 Radio Blocks (RB) have been shown in this Figure, but there are altogether (12+1)x4 = 52 blocks within a 52 TDMA frame) [Adapted from William Y.S.Lee[26], by permission from McGraw Hill].

The PDCCH comprises

(1) Packet Associated Control Channel (PACCH), a bidirectional channel used to pass signaling and other information between the MS and the network during packet transfer;
(2) Packet Timing Control channel (PTCCH), to control the timing advance for MSs.

(3) Packet Data Traffic Channel (PDTCH), to serve data transfer of actual user data for air interface in uplink or downlink.

The PRCH (also generally called Packet Data Channel – PDCH) uses a 52-multiframe structure (compared to 26 for GSM). Twelve (12) Radio Blocks (RBs), each consisting of 4 consecutive instances of a given slot, carry user data and signaling, as illustrated in Figure 9.1.

9.3. The B3G System

9.3.1. General

B3G stands for "Beyond 3G" and involves much higher data rate than 3G systems (which provides data rates only up to 2 Mbps). The ADSL and cable modem data speeds are much higher than the maximum data rate (2 Mbps) of 3G systems. Using different technologies than 3G, much higher data rates can be provided. Such technologies include the Wireless Loop Area Network (WLAN) which includes the Wi-Fi and WiMAX; and the resulting systems are referred to as B3G systems.

A series of standard specifications have been developed by the Institute of Electrical and Electronics Engineers (IEEE). These standards, referred to as IEEE 802.16, IEEE 802.11, IEEE 802.15, etc., apply to different cellular ("cell") radii and involve different data rates for various applications, as illustrated in Figure 9.2.

IEEE 802.11 operating in the 2.4-2.4835 GHz frequency band and providing data rates of either 1 Mbps or 2 Mbps, with spread spectrum transmission, represents an unlicensed band for industrial, scientific, and medial (ISM) applications.

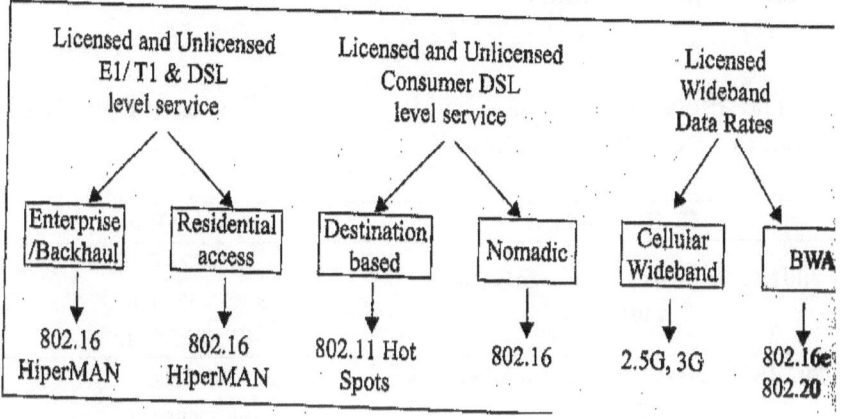

(a)

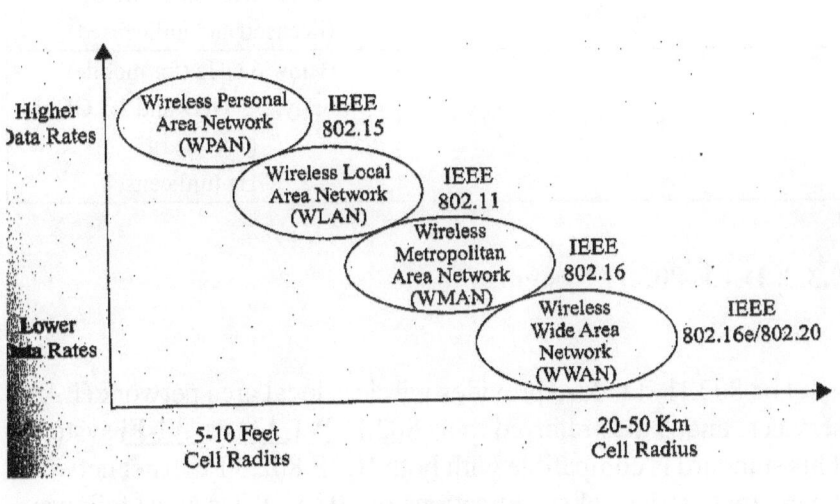

(b)

Figure 9.2.

(a) IEEE-based Wireless Standards versus Cellular Standards.
(b) Wireless Technology Comparison:

Wireless Area Network (WAN) versus IEEE Standards
(Adapted from William Y.S. Lee[26], ibid, p.296 and
297; by Permission from McGraw Hill).

9.3.2. Frequency Bands

The allocated frequency bands corresponding to various IEEE standards are indicated below.

IEEE Standard	Frequency Bands
802.11 Wireless Local Area Network (WLAN)	2.4-2.4825 GHz 5.15-5.35 GHz 5.47-5.825 GHz
802.15 Wireless Personal Area Network (WPAN)	2.4 GHz (unlicensed)
802.16 Broadband Wireless Access (BWA)	10-66 GHz (for LMDS) 2-11 GHz (for MMDS) (licensed and unlicensed)
802.16 e	Below 6 GHz (for mobile) [initially: 2.5 GHz and 3.5 GHz} (licensed)] 5.7 GHz (unlicensed)

9.3.3. IEEE 802.11 Systems

The 802.11 standard provides wireless local area network (LAN) services, and is also referred to as <u>802.11 WLAN</u> or **Wi-Fi** systems. This standard is compatible with both IEEE 802.3 Ethernet network. There are a variety of specifications of 802.11 WLAN, as follows:

Four different transmit transmission technologies are used in 802.11, viz., Infrared (IR), Frequency Hopping Spread Spectrum (FHSS), Direct Sequence Spread Spectrum (DSSS), and Orthogonal Frequency (OF). The pertinent four Standards are specified below.

(i) 802.11a – This standard operates at 5-GHz with an unlicensed band and provides transmission rates up to 54 Mbps. It is divided into four frequency bands, two of which require Dynamic Frequency Selection (DFS) and Transit Power Control (TPC). It uses Orthogonal Frequency Division Multiplexing (OFDM).

(ii) 802.11b – This specification operates in the 2.4 GHz band, providing data rates up to a maximum of 11 Mbps, over a maximum range of about 100 meters. Its uses Direct Sequence Spread Spectrum (DSSS).

(iii) 802.11g – This specification, an extension of 802.11b, also operates in the 2.4 GHz frequency band and the data rates provided are 1 Mbps and 2 Mbps, using Barker codes; and 5.5 Mbps and 11 Mbps using Complementary Code Keying (CCK)[27]. Higher data rates of 6, 9, 12, 18, 24, 36, 47, and 54 Mbps are also provided using Orthogonal Frequency Division Multiplexing (OFDM).

(iv) 802.11n – This specification provides data rates in excess of 100 Mbps, with OFDM most likely used in the Physical Layer. Advanced antenna techniques (e.g., MIMO and SIMO) are also employed. High throughput is achieved using this specification.

Other related specifications include 802.11d, 802.11e, 802.11f, 802.11h, 802.11i, 802.11j, and 802.11k. In particular, 802.11e includes consideration of Quality of Service (QOS); 802.11f of Inter-Access Point Protocol (IAPP); and 802.11i of security.

The frequency ranges and the overall hierarchy of the various 802.11 specifications are indicated in Figure 9.3.

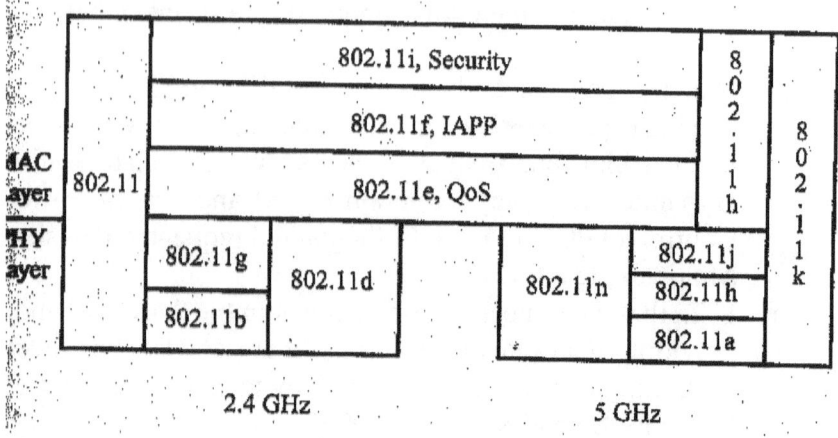

Figure 9.3. The hierarchy of 802.11 specifications.
(Adapted from W.Y.S. Lee[26], ibid, p.297; by Permission from McGraw Hill).

9.3.4. IEEE 802.11 Technologies

Some of the basic technologies employed in IEEE 802.11 are briefly described below.

9.3.4.1. Infrared (IR) Technology

The IR technology employed in mobile communication ranges in frequency from 300 GHz to 428,000 GHz. Typically, absolute line of sight (LOS) transmission (only) is used; since unlike radio propagation, which can penetrate walls and can be received even around corners, IR requires clear space for transmission. For this reason, IR signals are more secure and resistant to eavesdropping by a third party by interception of the signal secretly. 802.11 also provides additional security in the spread spectrum protocol, especially for protection against any adverse effect due to sunlight which can affect the IR transmission.

Usually pulse position modulation (PPM) is used synchronously for the IR signal, using short pulses with high peak power and low duty factor. The PPM also enhances security of the signal.

9.3.4.2. Direct Sequence Spread Spectrum (DSSS)

In DSSS, the narrowband signal is spread by directly multiplying it by a wideband pseudo-noise (PN) code sequence or by a smart code sequence known (only) to the transmitter and receiver both in advance. Typically, 1 MW to 100 MW power level is used in the frequency range of 2.4 GHz to 2.483 GHz over a time span of 1 Ms to 1 second.

For CDMA-DSSS, for each data bit to be transmitted, a redundant code bit pattern is generated, the code bit being called a chip. A longer chip pattern provides a higher probability for the original data bit to be recovered accurately, without the need of retransmission, since one or more bit in the chip pattern getting damaged does not cause failure. But obviously longer chip pattern reduces the data throughput, presenting a tradeoff. A narrowband unintended receiver without the PN code sequence or the smart code can intercept the transmission, but the transmission appears to it as low-power wideband noise, and therefore it ignores the transmission, being unable to detect any signal.

9.3.4.3. Frequency Hopping Spread Signal (FHSS)

In FHSS system, a narrowband carrier is used which hops over multiple carrier frequencies in a hopping sequence known (only) to the transmitter and receiver. A single logical channel is maintained by proper synchronization in FHSS. An unintended receiver without knowledge of the particular hopping sequence or pattern interprets the FHSS transmission as short-duration impulse noise. By virtue of its high security, FHSS is usually used in military communications.

9.3.4.4. Orthogonal Frequency Division Multiplexing (OFDM)

In the case of the OFDM 4-11 technology, a set of multi-carriers (subcarriers) obtained by subdividing the selected spectrum are used for transmission of high data rate in the presence of multipath fading. The subcarriers are used in the orthogonal mode; i.e., they are utilized independently, without any relation to one another. Thus, closely spaced subcarriers could be operated within a relatively small frequency spectrum, without the requirement of carrier guard-band overhead, as in the case of Frequency Division Multiplexing (FDM). A set of parallel bit stream can be transmitted using OFDM without suffering from inter-symbol interference (ISI). The ISI caused by time delay spread (TDS) is minimized further by the use of burst transmission mode, each burst carrying the data symbols preceded by a cyclic prefix large enough to absorb transients from previous bursts due to the multipath effect.

The use of the subcarrier technique reduces the symbol rate to be carried by each subcarrier. The symbol duration is, therefore, larger, which reduces the ISI under Rayleigh fading. Randomization of the burst error caused by wideband fading and frequency selective fading is also achieved in OFDM. The optimum transmission rate, coding rate, and modulation type is determined for the given channel conditions[27].

In a typical OFDM, eight data rates are used: 6, 9, 12, 18, 24, 36, 48, and 54 Mbps. The symbol period is $4\mu s$, composed of $3.2\ \mu s$ for data and $0.8\ \mu s$ for cyclic prefix. The lower the data rate, the higher the probability of correct reception. Out of a set of 54 subcarriers, 48 subcarriers are modulated by coded data symbols, while 4 subcarriers are used for pilot (or training) symbols. The subcarrier spacing is 311.5 kHz with 80 samples, corresponding to a sampling rate of 20 MHz. This provides a 64-point Fast Fourier Transform (FFT) or IFFT.

Different coding rates for convolutional coding are used depending on the level of protection required. Higher level of coding is required for a greater constellation of Quadrature Amplitude Modulation (QAM). OFDM with error correction codes are referred to as coded OFDM (COFDM).

9.3.4.5. Orthogonal Frequency Division Multiple Access (OFDMA)

The OFDMA technology refers to frequency division multiple access (FDMA) by users which share the orthogonal subcarrier(s) in the frequency domain. Subcarriers are divided into subchannels assigned to different subscriber stations. This allows multiple MSs to transmit simultaneously, thereby enhancing the network efficiency. Variable channel bandwidths can also be used, including 1.25 MHz, 3.5 MHz, 7 MHz, 14 MHz, 20 MHz, and 28 MHz.

9.3.4.6. IEEE 802.16-2004

The 802.16-2004 specification was introduced in 2004 to provide an alternative to wireless broadband access (e.g., cable DSL). This specification relates to Fixed Broadband Wireless Access (FBWA) and uses the frequency band of 10 GHz to 66 GHz (802.16c). This specification also relates to Local and Metropolitan Area Networks (LANs and MANs), and is also known as 802.16d. 802.16-2004 relates to a Fixed Wireless MAN (FWMAN), while 802.16 defines a mobile MAN (MMAN).

802.16-2004 specification includes four distinct physical layers (PHY) as shown in Figure 9.4, and is a point to multipoint (PMP) service, with traffic going through base station (BS); and traffic can also go directly between subscriber stations (SSs), in a mesh-network configuration.

Data Link Layer 2 ↕	Service Specific Convergence Sub-layer (SSCS)			
	MAC Common Part Sub-layer (CPS)			
Physical Layer 1 ↕	TDMA/TDM 10-66 GHz	Single Carrier 2-11 GHz	256 OFDM 2-11 GHz	2048 OFDMA 2-11 GHz

← Non-line-of-Sight (NLOS) Options →

Figure 9.4. 802.16-2004 Physical Layers
(Adapted from W.Y.S. Lee[26], Ibid, p.339; by Permission from McGraw Hill)

The 802.16-2004 data rates in a single carrier in the 10 GHz to 66 GHz frequency range are as shown in Table 9.1.

Table 9.1
802.16-2004 Data Rates

Channel Bandwidth	Bit Rate (Mbps)		
	QPSK	16 QAM	84 QAM
20 MHz	32	64	96
25 MHz	40	80	120
28 MHz	44.8	89.6	134.4

For the three PHY shown in Figure 9.4, power control/management and multiple antennas can be used to counter frequency selective fading (FSF) and multipath-induced delay spread problems. The single carrier TDM/TDMA physical layer is similar to the TDM/TDMA parameters as shown in Table 9.2.

Table 9.2
802.16-Single Carrier TDM/TDMA Physical Layer

Coding Rate	Modulation	Channelization Profile
1/2	BPSK	3.5 MHz
2/3	QPSK	7 MHz
3/4	16-QAM	10 MHz
5/6	64-QAM	20 MHz
7/8	256-QAM	

IEEE 802.16e: Mobile Wireless Metropolitan Area Network (MWMAN)

The 802.16e is designed to provide robust and efficient service operation under adverse mobile environment, with PHY layer based on OFDM/OFDMA; and is compatible with the fixed 802.16a. Table 9.3 provides a comparison of 802.16e with the two frequency band of 802.16-2004.

Forward Error Correction (FEC) and interleaving allow fine data granularity.

The Worldwide Interoperability for Microwave Access Forum (WiMAX Forum)[28-29] was formed in April 2001 in order to promote the adoption of IEEE 802.16-compliant equipment by operators of broadband wireless access systems (BWAX). WiMAX for wireless MAN (WMAN) can be compared to the Wi-Fi Alliance that aims to promote the adoption of the IEEE 802.11 standard for wireless LANs (WLANs).

Table 9.3
Comparison of 802.16e versus 802.16-2004 Standards

	802.16e	802.16-2004 2-11 GHz Band	802.16-2004 10-66 GHz Band
Date of Standard Completion	2005	June 2004	June 2004
Channelization	Variable 1.5-20 MHz Uplink Subcarrier	Variable 1.5-20 MHz	20, 25, and 28 MHz
Cell Radius	4-6 miles	4-6 miles	1-3 miles
Data Rate	1.5 Mbps downlink 15 Mbps uplink in 5 MHz channel	1 to 70 Mbps in 14 MHz channel	32 to 134.4 Mbps in 28 MHz channel
Access Technique	OFDMA	Single Carrier 256 OFDM, 2-K OFDMA	Single Carrier
Modulation	QPSK, 16 QAM, 64 QAM	QPSK, 16 QAM, 64 QAM	QPSK, 16 QAM, 64 QAM
Mobility	12 to 60+ mph	None	None
Line of Sight (LOS)	No	No	Yes
Licensed and Unlicensed	Licensed only	Yes	Yes

9.3.4.7. Wireless-Fidelity (Wi-Fi)

The Wi-Fi Alliance (www.wi-fi.org) is a nonprofit international association formed in 1999 to certify compatibility and interoperability of wireless LAN equipment and products based on IEEE 802.11 specifications. The certification process began in 2000; and so far more than 1500 products have been certified, and 200 companies from around the world have become members of the Wi-Fi Alliance. The current certification includes Wi-Fi Multimedia (WMM), Wi-Fi

Protected Access 2 (WPA2), and 11a, 11d, and 11g. A number of labs in many countries perform testing for certification.

9.3.4.8. Hot Spot[30]

A public *hot spot* refers to a publicly available wireless network connection location where connections with other users-compatible wireless network devices, such as cell phones, notebook computers, hand-held communications and games, PDAs, etc., may be readily connected. Such users can connect to the Internet or private Intranet(s). Thus they can send or receive e-mail, or upload or download files in the wireless mode. Many major wireless cellular or mobile service providers make such hot spots available for the public.

The unlicensed IEEE 802.11 is commonly used for providing hot spots. The companies that are known as wireless Internet service providers (WISP), have been proliferating rapidly in recent years. The wireless LAN (WLAN) offering access to the public is referred to as PWLAN.

9.3.4.9. 3G Networks

The main features of the 3G system (also called IMT 2000) development over what was called by ITU the "Future Public Land Mobile Telephone System" (FPLMTS) are as follows:

- Spectrum bandwidth 5 MHz
- Wideband CDMA modulation and multiple access
- High spectrum efficiency (compared to 2G system)
- Full coverage and mobility for data rates 144-384 kbps
- Limited coverage and mobility (or no mobility) for 2 Mbps
- Greater flexibility toward introducing new and multimedia services
- Packet data mode operation

- Provision of multi-rate services
- Efficient power control
- Inter-carrier handover
- A user-dedicated pilot channel for a coherent uplink
- Beam forming using a downlink pilot channel
- Multiuser detection scheme
- Circuit-switching (CS) and packet-switching (PS)
- Mobile vehicle service at 144 kbps
- Pedestrian user service at 384 kbps
- Fixed/indoor user and pico-cell service at 2 Mbps
- Additional flexibility for future system for making hooks and extension to make intersystem connections among three systems:

    (1) cdma2000 in North America
    (2) WCDMA-ULTRA in Europe
    (3) WCDMA-ARIB in Japan

Additional features of the above three systems are depicted in Table 9.4.

Table 9.4
Some features of the three IMT 2000 systems (FDD – Frequency Division Duplex; TDD – Time Division Duplex; CDM – Code Division Multiplexing; TDM – Time Division Multiplexing)

	FDD DIRECT SPREAD	FDD Multicarrier	TDD
Bandwidth (MHz)	5	5/1.25	5/1.6
Chip Rate (Mcps)	3.84	3.6864/1.228	3.84/1.28
Synchronization	Asynchronous/synchronous	Synchronous	Synchronous
Core Network	GSM-MAP/IP	ANSI-41/IP	GSM-MAP/IP
Common Pilot	CDM	CDM	TDM
Dedicated Pilot	TDM	CDM	TDM

CDMA2000 3G System for North America

The IS-95 air interface specifications were extrapolated to define the CDMA-One system, which were then made to meet the requirement for IMT-2000 to generate the CDMA-2000 system (a 3G system) for North America.[31-34]. This Third Generation Cellular System is backward compatible with the IS-95.

CDMA-2000 is implemented using both signal-carrier and multiple-carrier; and the physical layer channels for both Frequency Division Duplex (FDD) and Time Division Duplex (TDD), with FDD to be implemented first. The physical layer comprises both dedicated physical channels (DPHCH) and common physical channels (CPHCH), for forward and reverse communications including control channel and pilot channel.

CDMA-2000 FDD Radio Interface

The radio interface of CDMA-2000 is similar to that of IS-95. Thus, after a CDMA-2000 frame is generated by the physical channel, the physical layer adds the CRC bits for detecting frame errors, provides the FEC coding bits, and performs interleaving to counter the long-term fading.

The chip rates used by CDMA-2000 are in the range given by a multiple of 1.2288 Mbps, i.e., by N x 1.2288 Mbps, $N = 1, 3, 6, 9, 11$.

For $N > 1$, CDMA2000 spreads the signal in two ways:

(1) Multicarrier – Demultiplexes the message signal into N information signals and spreads each of those on a different carrier at a chip rate of 1.2288 Mbps, in consistence with the IS-95 signal format.
(2) Direct spread – in this case the message signal is spread directly with a chip rate of 1.2288 x N Mcps.

The <u>pulse waveform</u> for the transmission conforms to IS-95.

The <u>Frame-structure</u> of CDMA2000 entails two frame-lengths: 5ms and 20ms. The 20ms frame-length is used to enhance the performance of the demodulator by employing longer interleaving span.

The <u>Modulation chip rate</u> is 1.2288 Mbps x N, and the <u>modulation technique</u> is as follows:

<u>Forward Link</u>
QPSK
(for both data and spreading modulation)

<u>Reverse Link</u>
BPSK for data modulation
QPSK for spreading modulation

The <u>Channel Spacing</u> is $(N + 1)$ x 1.25 MHz.

The <u>transmission characteristics for CDMA2000 TDD</u>[35-37] are the same for FDD; but the transmission takes place in TDMA burst mode, the guard times are introduced appropriately, and time division multiplexing (TDM) is used. On the forward link, each time slot of 1.67ms contains 2048 chips, corresponding to a net chip rate of 1.2288 Mcps (= 2048 chips/1.67ms). There are 16 time slots in the TDM chip stream.

9.3.4.10. Other Networks

<u>1xEV-DV Network</u>

The high-speed enhancement of CDMA2000 Revision C and Revision D is called 1xEV-DV system[35-37]. The bandwidth efficiency of the CDMA2000 system is enhanced by the introduction of the Forward Packet Data Channel (F-PDCH) and the Reverse Packet Data Channel (R-PDCH). These data channels use fast packet scheduling, hybrid automatic repeat request (HARQ), and adaptive modulation and coding (AMC) for this purpose. AMC uses power control in the R-PDCH to counter time varying multipath fading.

1xEV-DV sends both voice and data signals using flexible TDM-CDM multiplexing, so they share the same carrier of 1.25 MHz spectrum in the VoIP packet stream. The uplink channel data rate is 3.1 Mbps. This system can provide a two-way video conference with live image call at 15-frames per second.

1xEV-DO Network

1xEV-DO, using different protocols than those of CDMA2000, uses a seven-layer protocol architecture for connectivity between packet switched data network (PSDN) and access terminal (AT). The seven layers are listed below:

(1) Application Layer
(2) Stream Layer
(3) Session Layer
(4) Connection Layer
(5) Security Layer
(6) MAC Layer
(7) Physical Layer

1xEV-DO represents enhanced version of single carrier (1x) for data only and operates in two modes: single mode and dual mode. The single mode operation involves accessing the high-speed data rate (HDR) services at 2.4576 Mbps only for radio channels. The Dual Mode can access this (EVDO HDR) channel or IS-95 voice with medium-rate data traffic channels. Data connectivity to the MS is also provided with the help of the access terminal (AT).

9.4. The 4G System

9.4.1. Introduction

The 4G Systems in cellular and mobile wireless services reflect technological advances in many related areas, including the following notable developments among others.

(a) Hybrid networks – Hybrid implementation of broadband and cellular networks provide higher spectral efficiency. In such networks, high bandwidth downlink multicast (destined to a multiplicity of destinations) data over a broadband are combined with cellular services.
(b) Optimized Transmission – The modulation scheme and coding power control, in conjunction with the so-called 'software defined radio (SDR),' applied with adaptive and reconfigurable systems yield high grade of service under varying transmission conditions. In particular, the low-density parity check codes (LDPC), the turbo coding technique and high-order modulation yield a high level of spectral efficiency as well as flexibility in implementation.
(c) Multiple Input Multiple Output (MIMO) – The use of the MIMO technique, in combination with multiple antenna array (MAA) technology, allows enhancement in the system efficiency.
(d) Orthogonal Frequency Division Multiple Access (OFDMA) – The OFDMA approach based on the use of orthogonal frequencies for multiple access in the frequency domain multiplies the system capacity, characterized by the number of frequency reuse method within the network.
(e) Multiple Carrier Code Division Multiple Access (MC-CDMA) – MC-CDMA allows capacity enhancement with the use of multiple (rather than a single) carrier used in the CDMA mode.
(f) Single Scheduling Packet-Based System (SSPS) – SSPS is another technique useful toward enhancement of the system capacity.
(g) Advanced Antenna Technology – More efficient antenna design is used, for example, antenna arrays serve to provide greater antenna gain while reducing inter-cell interference; thereby adding to the system capacity significantly.

The combination of OFDMA and MIMO techniques are widely employed in the 4G system. The 4G system has been pursued by many international wireless companies, with a number of innovative implementation approaches.

South Korea hosted the Second 4G Forum on August 24-25, 2006, in which a number of international companies[38-43] presented their ideas and proposals.

Mobile data transmission usually involves low data speeds, and differentiating between networks and options is quite straightforward. Data speeds mainly depend on the coverage area as well as available bandwidth on the network. 3G speeds varied around between 500 kbps to the 2 Mbps range, and design of modem or mobile hot spot meeting desired specifications was relatively simple.

In the case of the 4G data system, the International Telecommunications Union (ITU) defines it as a connection capable of 100 Mbps with high mobility, and up to 1Gbps with low mobility (Wi-Fi range). The US carriers aim at providing 4G data networks; though no carrier actually meets the ITU official definition. The cellular data network's 4G speeds remain well below the official requirements; and the only definition each network seems to be able to agree upon is that 4G is just what comes after 3G. While the title of "4G" generally meaning the *4th Generation* isn't necessarily accurate, nonetheless this terminology is commonly used.

A Quick Look at "4G" Technologies

It is useful to take a quick look at the different technologies that are currently being labeled as "4G" in the United States.

- Mobile WiMax - WiMax is the "4G" technology that some carriers use, and it offers peak data rates of 128 Mbps downstream and 56 Mbps upstream.

- Long Term Evolution (LTE) - LTE is an alternative choice for "4G" mobile broadband, providing theoretical peak data rates of 100 Mbps downstream and 50 Mbps upstream. While LTE (or, specifically, 3GPP LTE) isn't technically 4G, LTE Advanced is expected to actually meet 4G requirements with a peak download speed of 1Gbps. The upgrade path from

3GPP LTE to LTE Advanced is supposed to be easier and more cost-effective than most upgrades.
- High-Speed Packet Access (HSPA) – This is still another variation of 4G, even though HSPA is what some providers use for its 3G data. While HSPA offers faster speeds, those peak speeds are about half of what LTE and WiMax offer—56 Mbps downstream and 22 Mbps upstream.

Clearly different carriers have taken fairly different approaches to what they're calling their 4G networks. These choices make for very different strengths and weaknesses in each, primarily in the categories of coverage, speed, device options, and operating system support.

References:

[26]   William C.Y. Lee, Wireless and Cellular Telecommunications, McGraw Hill, 2006.

[Note-It must be thankfully acknowledged that Reference 26 (W.C.Y.Lee) served to the author (A.K.Sinha) of this book as a primary Reference in the preparation of many segments of this book. A number of Figures, data, and Tables have been used in the present book from Reference 26, by Permission from the Publisher: McGraw Hill, as mentioned appropriately in the text of this book].

[27]   S. Hartford, M. Webster, and J. Zyren, "CCK-OFDM Normative Text Summary," IEEE 802.11-01/436 rl, July 2001.

[28]   Intel White Paper, "IEEE 802.16 and WiMAX Broadband Wireless Acess for Everyone," http://grouper.ieee.org/groups/802/16.

[29]   C.Eklund, R.B. Marks, K.I. Stanwood, and S. Wang, "IEEE Standard 802.16: A Technical Overview," IEEE Communication Magazine, June 2002.

[30]   Daniel Minoll, Hot Spot Networks, Wi-Fi for Public Access Locations, McGraw Hill, 2002.

[31] Daniel Minoll, Hot Spot Networks, Wi-Fi for Public Access Locations, McGraw Hill, 2002.
[32] T. Ojanpera and R. Pasad, Wideband CDMA for Third Generation Mobile Communications, Artech House Publishers, Boston, 1998 [Chapter 5 CDMA Air Interface Design].
[33] V.K. Garg, IS-95 CDMA and cdma2000, Prentice Hall, PTR, 2000.
[34] V. Vangi, A. Damnjanovic, and B. Vojcic, The cdma2000 System for Mobile Communications, Prentice Hall PTR, 2004.
[35] S.C. Yang, 3G cdma2000, Artech House, Boston, 2004.
[36] B. Pelletier and H. Leib, "PCS Third Generation CDMA System, Study of the Physical Layer," Wireless Communication Group at McGill University, Canada, August 2004.
[37] H. Holma and Antti Toskala, WCDMA for UMTS, John Wiley & Sons, 2001.
[38] C. Smith and D. Collins, 3G Wireless Works, McGraw Hill, 2002.
[39] J.S. Lee and L.E. Miller, "CDMA System Engineering Handbook," Artech House, Boston, 1998.
[40] 3GPP2, "cdma2000 Standard for Spread Spectrum Systems, Revision C, May 2002.
[41] 3GPP2, "cdma2000 Standard for Spread Spectrum Systems, Revision D," February 2004.
[42] The participating companies in this Forum included Samsung (Host), Nokia, Motorola, Ericson, Siemens, Wireless Broadband (WiBro), NTT McCoMo, etc.
[43] W.C.Y. Lee, CS-OFDMA: A New Wireless CDD Physical Layer Scheme, IEEE Communication Magazine, Vol. 43, February 2005, p.74-49.

10. TYPES OF SIGNALS

10.1. Introduction

Advantages of Digital Communications

Advantages of digital telecommunications have been indicated before in this book; and are reiterated here again in view of their basic implicit significance and of the fact that the present and future telecommunications applications are likely to be exclusively digital in nature. Digital communication techniques are relatively far more advantageous in several ways, including:

- Possibility of integrating different service applications such as voice, video, data, etc., is enhanced;
- Higher bandwidth efficiency is generally achievable;
- Noise reduction through bit-regeneration permits a greater level of signal quality;
- Cost reduction is realized by using Large-scale Integrated Circuit (LIC) and microprocessing technology permitting mass-production of systems and components (e.g., digital switching) through economy of scale;
- The signals can be manipulated with high degree of flexibility thereby allowing new innovative service applications and devices;
- Advanced coding and encryption is readily and economically available, affording a greater level of security of the signal;
- The cost to the end-user usually significantly reduced;
- Remote location and mobile services are easy to implement.

For the above reasons, digital transmissions and systems have almost completely replaced the analog systems for various service applications during the last few decades.

In this book, most of the material pertains to digital communications due to the current prevalence of digital technology and applications. Here we recapitulate various important aspects involved in digital telecommunications as applied to various service applications: digital voice,

10.2. Digital Voice

10.2.1. General

Human voice signals typically occupy a bandwidth of 4 kHz, and hence a bandpass filter operating in the range of 300-3400 Hz is employed for voice communications.

As a sampling rate of at least equal to twice the highest frequency in the input baseband signal is basically required, the sampling rate for voice signal is usually 8,000 samples per second, since the maximum frequency entailed in voice is 4 kHz. Without any bit reduction technique, each sample could be encoded by using 8 bits/sample. This leads to a bit rate of 64 kbit/sec (also written as kbps) for voice without any special processing or bit-reduction. However, in practice the actual bit rate employed may vary over a wide range depending on the specific application, as indicated below:

Synthetic-quality voice – 2.4-4.8 kbps
Private voice networks – 4.8-32 kbps
Local and long-distance telephony networks – 8-32 kbps
Broadcast-quality and Hi-Fi audio – 64-400 kbps

The system efficiency is of course dependent on the modulation, multiple-access, coding, channel-noise, among other factors involved. Some of the major techniques pertaining to these factors are mentioned below to summarize an augment the relevant factors already discussed in this book earlier).

Modulation

- PCM (Pulse Code Modulation)
- PAM (Pulse Amplitude Modulation)
- PPM (Pulse Position Modulation)
- DPCM (Differential PCM)
- ADPCM (Adaptive DPCM)
- PSK (Phase Shift Keying)
- BPSK (Binary PSK)
- QPSK (Quaternary PSK)

Multiple Access (MA)

- FDMA (Frequency Division MA)
- TDMA (Time Division MA)
- SCPC (Single-Channel-Per-Carrier)

Multiplexing (MUX)

- Frequency Division MUX (FDM)
- Time Division MUX (TDM)

Coding

- Block coding
- Convolution coding

Noise-Type

- Quantizing noise
- Thermal noise
- Interference noise

- Bandwidth reduction or truncation noise
- Intermodulation noise

10.2.2. Signal-to-Noise Ratio (S/N) for Quantization Noise

10.2.2.1. Signal to Quantization Noise Ratio

A critical parameter in the signal transmission is the ratio of the signal power to the noise level due to various sources of noise. In a digital system, a basic measure of the received signal quality is the ratio of signal power to quantizing noise, since some form of quantization is always used for transmitting voice or audio signal which is originally an analog signal (varying continuously within a certain overall range), the sampled value of the instantaneous amplitude is necessarily represented only *approximately* through a set of quantum levels, and hence the reproduction of the signal at the receive end contains the resulting 'quantization noise.'

If the original (analog) signal is coded by using N quantum levels, and coded using n bits/sample, then since each bit can take two possible values (0 or 1), n bits can represent a total of 2^n levels, so that

$$N = 2^n \qquad (10.1a)$$

The quantization noise generally varies as the square of the separation between successive quantum levels; so for N levels which correspond to $(N - 1)$ inter-level separations, we can write:

$$(N - 1)l = A$$

i.e.,

$$l = A/(N - 1) \qquad (10.1b)$$

where l is the mean or constant separation between successive levels and A is the amplitude of a sinusoidal wave-form that is an envelope of the varying signal. The quantization noise can be written as:

$$N_q = C_1 l^2 \qquad (10.1c)$$

where C_1 is a constant.

The (maximum) signal power S_m is proportional to the square of the amplitude A, i.e.:

$$S_m = C_2 A^2 \qquad (10.1d)$$

where C_2 is a constant.

Combining Equations (10.1a-d), we have the ratio of the signal power to quantization noise given as:

$$S_m/N_q = \frac{C_2 A^2}{C_1 l^2} = \left(\frac{C_2}{C_1}\right)(N-1)^2$$

$$= C_0(2^n - 1)^2 \qquad (10.2a)$$

where $C_0 = C_2/C_1$ is also a constant. Since $2^n \gg 1$ for most practical choices of n (e.g., for $n = 8$, $2^n = 256 \gg 1$), the above equation can be approximated as

$$S_m/N_q = C_0(2^n)^2 = C_0 2^{2n} \qquad (10.2b)$$

Tests and experimental observations typically yield the value

$$C_0 \simeq 3/2 \qquad (10.2c)$$

Consequently, we can write Equation (10.2b) as

$$S_m/N_q \simeq (3/2)\, 2^{2n} \qquad (10.2d)$$

It may be noted that since a test-tone in telephony is usually $3dB$ below (i.e., half) the signal level, the corresponding ratio for a test-tone can be simply written as:

$$S_t/N_q = (3/4)\, 2^{2n} \qquad (10.2e)$$

The quantization may be uniform (linear), or non-uniform (nonlinear), with the inter-level separation progressively increasing for higher amplitude range, since greater fidelity is required as the signal level decreases. Two types of variation patterns of the quantization levels have been adapted as standards in North America and Europe, called the μ-law and A-law, respectively. Thus, if we represent the input signal level as x normalized to unity (1), and the sampled quantum value as y, then the above two standard laws are expressed by the equations given below.

10.2.2.2. The μ-Law (North American Standard)

$$y = \frac{\ln(1+\mu x)}{\ln(1+\mu)} \qquad (10.3a)$$

where $\ln$ denotes natural logarithm (with respect to base e), and $\mu = 255$.

10.2.2.3. The A-Law (European Standard)

$$y = \frac{Ax}{1+\ln(A)},\; 0 \le x \le \frac{1}{A} \qquad (10.3b)$$

$$y = \frac{1+\ln(Ax)}{1+\ln(A)},\; \frac{1}{A} < x < 1 \qquad (10.3c)$$

where $A = 876$ for this standard.

Linear or uniform quantization is not very efficient except in the case of high level of input signal; and a non-uniform (nonlinear) quantization level pattern, with the help of a companding process is commonly used.

The variation pattern of the signal-to-quantization noise ratio as a function of the input signal level (amplitude) for uniform (linear) and for companded PCM was provided by the Bell Telephone Labs. (Transmission Systems for Communication, Fourth Edition, 1970). For uniformly quantized PCM (i.e., without companding) this ratio decreases decibel for decibel with decreasing input signal level; but for companded PCM system this ratio remains nearly constant for input signal level above about -25 dB.

10.2.3. Signal-to-Noise Ratio (S/N) for Thermal Noise

Thermal Noise

The thermal noise occurs due to random motion of electrons in the communication equipment. Therefore, the thermal noise is proportional to the number of electrons per unit bandwidth and obviously, is also a function of the so-called *noise temperature* (T) characterizing the thermal level (room-temperature, cryogenically cooled system, etc.) of the system.

According to the well-known Fermi-Dirac formula, the number of electrons within the frequency interval v to $v + dv$ is given by:

$$f(v)dv = \frac{hv}{1-e^{-hv/kT}} \qquad (10.4a)$$

where h is the Planck constant ($h = 6.626 \times 10^{-27} \, erg.sec$), and k is Boltzmann constant ($k = 1.381 \times 10^{-16} \, erg/K$).

For noise temperature large enough so that $hv \ll kT$, the approximation series expansion can be made

$$e^{-hv/kT} = 1 - \frac{hv}{kT} + \frac{1}{2}\left(\frac{hv}{kT}\right)^2 - \frac{1}{3}\left(\frac{hv}{kT}\right)^3 + \cdots \text{ad infinitum}$$

(10.4b)

For high enough temperature

$$kT \gg hv \qquad (10.4c)$$

we can neglect the second and higher order terms on the right of Equation (10.4b), so that in Equation (10.4a), we can simply write

$$e^{-hv/kT} \simeq 1 - \frac{hv}{kT} \qquad (10.4d)$$

and also set $dv = 1$ Hz for unit bandwidth. Then, combining Equations (10.4a) and (10.4d), we can write

$$N_o = kT \qquad (10.5a)$$

where N_o represents the energy density (i.e., energy per unit bandwidth) due to the thermal noise.

Due to the thermal noise, one or more bits in the coded message may undergo bit error. The amount of signal degradation depends on the position(s) of the bit(s) that suffer bit error(s). If a corruption (bit error) occurs in the least significant bit (LSB) of a PCM code word, one quantum level will become erroneous in the received message code word, and this corresponds to an erroneous determination by one quantum level, amounting to N. A bit error in a bit which is the next higher significant bit will result in an error equivalent to double the amount (i.e., $2N$), the next higher significant bit, if it undergoes bit error, will result in an error corresponding to $4N$ in amount, and so on. It can be shown[44] that the signal-to-noise induced bit errors can be written as

$$\left(\frac{S}{N}\right)_t = \frac{1}{4P_e} \qquad (10.5b)$$

where P_e is bit error rate (BER), i.e., the probability of one bit error, given by

$$P_e = \frac{1}{2}\left[erfc\left\{\left(\frac{E_b}{N_o}\right)^{1/2}\right\}\right] \qquad (10.6a)$$

In Equation (10.6a) above, $erfc$ represents the integral

$$erfc(x) = \frac{2}{\sqrt{\pi}} \int_x^\infty e^{-u^2} du \qquad (10.6b)$$

and is called the *complementary error function*. The numerical value of *erfc* function as a function of x is available for reference in tabular form in many books.

If quantization noise and the degradation due to bit error resulting from other sources of noise (thermal and channel) are considered together, the signal-to-noise ratio due to total or combined noise is given as

$$\left(\frac{S}{N}\right)_T = \frac{C_B}{1+4C_B P_e} \qquad (10.6c)$$

where

$$C_B = C_o 4^B \tag{10.6d}$$

and C_o is a constant whose value depends on *instantaneous companding* (IC).

For low values of P_e, the performance is dominated by quantization noise, so that the $\left(\frac{S}{N}\right)_T$ remains nearly constant (i.e., no improvement) with increasing values of carrier-to-noise ratio (C/N). However, in the lower range of the (C/N)-value, as the (C/N)-value is increased, the bit error rate (P_e) decreases (BER-dominated region), and the thermal-related $\left(\frac{S}{N}\right)_t$ (Equation 10.5b), as well as the total $\left(\frac{S}{N}\right)_T$ (Equation 10.6c) increases.

10.2.4. Phase Shift Keying (PSK)

In case of digital transmission with phase shift keying (PSK) modulation, the carrier phase is modulated by the input bit-stream by altering the phase of the carrier appropriately. Different common modes of PSK modulation include BPSK, QPSK, 8ϕ-PSK, or QAM, as outlined before. A few additional pertinent factors pertaining to BER and performance are discussed below.

10.2.4.1. Binary PSK (BPSK)

In the BPSK digital modulation, only two possible choices of the phase (ϕ) variation is incorporated, viz. $\phi = 0 \; or \; \phi = \pi$ (radians), corresponding to the bit being 1 or 0, respectively.

Since a single bit, with two possible values (0 and 1) suffices to represent the stated two-phase state, each sample of a BPSK-modulation contains a single bit. In the phase-space, the BPSK is simply represented by a horizontal straight line through the origin, with its left arm representing the phase $\phi = \pi$ (i.e., bit =0), and the right arm representing the phase $\phi = 0$ (i.e., bit = 1), as illustrated in Figure (10.1).

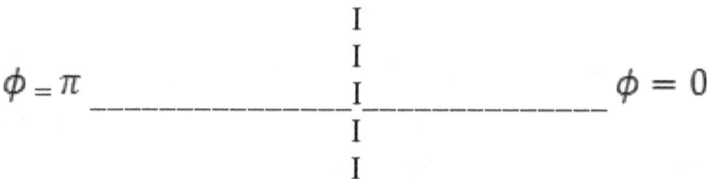

Figure 10.1: BPSK phase diagram.

It is possible to employ a carrier orthogonal to the stated carrier, such that the phase $\phi = \pi/2$ represents the bit 1 and the phase $\phi = \frac{3}{2}\pi$ represents the bit 0, as shown by the vertical line in Figure (10.1). The two orthogonal carriers can be operated independently with respect to each other. For distinguishing the two carriers, the one represented by the horizontal line is referred to as the i-channel, while the orthogonal carrier represented by the vertical line is referred to as the q-channel.

10.2.4.2. Offset Quarternary PSK (OQPSK)

In QPSK modulation, each sample contains two bits. Since each of these two bits can represent two possible states (for bit = 0 and bit = 1), the total number of states represented by two-bit sample is $2 \times 2 = 2^2 = 4$. In the phase-space, these four states are represented in the offset-QPSK (OQPSK) system by the four possible phase values, viz. $\phi = \pi/4$; $\phi = \frac{3}{4}\pi$; $\phi = \frac{5}{4}\pi$; and $\phi = \frac{7}{4}\pi$

, respectively. The four possible bit combinations for two-bit sample corresponding to the above four states could be taken as for instance,

$$(1\ 1) \rightarrow \text{corresponding to} = \pi/4,$$
$$(0\ 1) \rightarrow \text{corresponding to } \phi = \frac{3}{4}\pi,$$
$$(0\ 0) \rightarrow \text{corresponding to } \phi = \frac{5}{4}\pi,$$
$$\text{and } (1\ 0) \rightarrow \text{corresponding to } \phi = \frac{7}{4}\pi.$$

10.2.4.3. Eight-Phase PSK (8ϕPSK)

By a simple extension of the BPSK and QPSK, the 8ϕPSK is defined as the case when each symbol comprises three bits. The possible number of states in this case is therefore, $2^3 = 8$. A possible representation of 8ϕPSK in the phase-space is thus possible corresponding to the following choice of combinations of the three-bit sample and phase (ϕ)-values (Table 10.1):

Table 10.1
Possible Choice of Bit-Sequences for an 8ϕPSK

3-bit Combination of a single 8ϕPSK sample	Phase (ϕ) Value (radians)
(0 0 0)	$(1/6)\pi$
(0 0 1)	$(1/3)\pi$
(0 1 0)	$(2/3)\pi$

(0 1 1)	$(5/6)\pi$
(1 0 0)	$(7/6)\pi$
(1 0 1)	$(4/3)\pi$
(1 1 0)	$(5/3)\pi$
(1 1 1)	$(11/6)\pi$

It may also be noted that the frequency spectra of both BPSK and QPSK are of the form:

$$W(w) = \left(\frac{\sin w}{w}\right)^2 \qquad (10.6e)$$

Performance

For a bandwidth B, the total thermal noise (N_t) is given by

$$N_t = N_o B = kTB \qquad (10.7a)$$

where $N_o = kT$ is the noise power density for thermal noise for a carrier power C, the carrier to noise ratio, C/N, can therefore be written as

$$C/N = C/N_o B \qquad (10.7b)$$

If the bit rate number of bits per second is R_b, i.e.,

$$R_b = 1/T_b \tag{10.7c}$$

where T_b is the bit period (in seconds), then clearly,

$$C = E_b/T_b \tag{10.7d}$$

Combining Equations (10.7b-d), we have the result:

$$\frac{C}{N} = \frac{E_b R_b}{N_o B} = \left(\frac{E_b}{N_o}\right)\frac{R_b}{B} \tag{10.7e}$$

The ratio (E_b/N_o), referred to as the *energy-per-bit-to-noise-density ratio*, is an important basic parameter in digital communications. The graphs of bit error rate (BER, i.e., P_e) versus (E_b/N_o) for a particular type of modulation (e.g., BPSK, QPSK) is an important piece of information on which to determine the transmission system design based on the minimum required level of performance for a specific service application, or for comparison between the theoretically expected performance and the test or measured data. The uplink and downlink system parameters (antenna size, carrier frequency, transmitter power, system noise temperature, satellite repeater-gain, etc.) and related trade-offs are carefully adjusted and utilized in view of the given noise environment for a satellite link, for example, to ensure that, in order not to exceed the maximum allowable BER requirements for a particular service application, translated into the minimum (E_b/N_o) requirement is properly adhered to. General characteristic maximum BER values for a few common applications is shown in Table 10.2. Note that the values in Table 10.2 (for early 90s)[45] may be now superseded under new technology.

Table 10.2. Typical Performance Requirements for a few Digital Service Applications. (from Pritchard et al, *ibid*, p.347)[45]

Service Application	Bit Rate (R_b) in kbps	Maximum Allowable BER	Typical Connect Time (minutes)
Digital Voice	19.2-64	10^{-4}	3-4
Business Video	56-1544	10^{-5}	30-60
Data File Transfer	56-6312	10^{-8}	2-30
Electronic Mail, Fax	4.8-56	10^{-8}	2-10
File Downloading	9.6-56	10^{-8}	2-10
CAD-CAM	56-224	10^{-7}	10-60
Remote Job Entry	9.6-56	10^{-8}	10-30
Computer Graphics	9.6-56	10^{-8}	30-60

10.3. Bit Error Rate (BER) for Phase Shift Keying (PSK) Modulation

The variation pattern of BER as a function of E_b/N_o depends on the choice of the type of PSK. The variation pattern for BPSK, QPSK, and 16-ϕ (16-phase)-PSK are briefly discussed below.

10.3.1. BPSK

It may be noted that for BPSK modulation, the required bandwidth (B) is typically 1.1 to 1.2 times the bit rate (R_b). Theoretical and measured variation patterns for BER versus E_b/N_o, below BER =

10^{-7} show that each increase by 1 dB in E_b/N_o typically reduce the BER by one order-of-magnitude (i.e., by a factor of 10).

10.3.2. QPSK

Recall that the QPSK uses half the bandwidth as compared to the BPSK, and the sample rate in QPSK (2 bits/sample) is the same as the bit rate for the BPSK (1 bit/sample). Thus, the bit rate for QPSK (i.e., the number of bits per second) is double that of BPSK. Consequently, the E_b/N_o values for the two modulation schemes for a given noise environment can be related as follows:

$$(C/N)_B = \left(\frac{E_b R_b}{N_o B}\right)_B \qquad (10.8a)$$

where the suffix B refers to BPSK. Replacing R_b by $2R_b$, and B by $(1/2)B$ while keeping the N_o constant, and using the suffix Q to refer to QPSK, and also noting that $(E_b)_Q = \frac{1}{4}(E_b)_B$ for a constant carrier power (C), since

$$(E_b R_b)_B = (C)_B = (C)_Q = (2E_b)_Q \times 2(R_b)$$

i.e., $(E_b)_Q = \frac{1}{4}(E_b)_B$

we obtain

$$(C/N)_Q = \left(\frac{E_b R_b}{N_o B}\right)_Q = \frac{1}{4}E_b \times (2R_b)/N_o \times \left(\frac{1}{2}B\right)$$

$$= \left(\frac{E_b R_b}{N_o B}\right)_B \qquad (10.8b)$$

In other words, the E_b/N_o values for the two cases remain the same for identical conditions. Thus the same BER versus E_b/N_o theoretical curve applies for both BPSK and QPSK:

$$(E_b/N_o)_B = (E_b/N_o)_Q \qquad (10.8c)$$

The theoretical (and measured) performance curves for BPSK and QPSK, therefore remain identical.

10.3.3. 8-Phase PSK (8ϕPSK)

In case of 8ϕPSK, the concept of Gray code bit mapping (GCBM) is useful. GCBM refers to the condition that adjacent samples, each consisting of 3 bits, differ by only 1 bit. Under this and certain other simplifying assumptions (symmetric decision boundaries), the BER or the bit error probability (P_e) can be expressed as follows[46]:

$$P_e\vert_{8\phi} = \frac{3}{2}\left[f\left(\frac{13\pi}{8}\right) - F\left(\frac{\pi}{8}\right)\right] \qquad (10.9a)$$

where the function $F(\phi)$ with phase (ϕ) as a variable can be written as

$$F(\phi) = \frac{-\sin\phi}{4\pi}\left[\int_{-\pi/2}^{\pi/2} \frac{\exp(-6\,E_b/N_o)(1-\cos\phi\cos\xi)}{(1-\cos\phi\sin\xi)}\,d\xi\right]$$

$$(10.9b)$$

The 8ϕPSK is useful for mobile communications due to its bandwidth efficiency, critical for accommodating a large number of (mobile) users within a given bandwidth, and due to high and varying noise environment.

The GCBM criterion mentioned above, however, allows the assumption that the bit error rate (P_e) is equal to the symbol error rate (probability), P_s, in which case the above exact but complicate formula for 8ϕPSK (Equations 10.9a and 10.9b) become simplified, at least for an approximate result:

$$P_e I_{16} \simeq \frac{1}{3} \, erfc\left[\left(\frac{3E_b}{N_o}\right)^{1/2} \sin\left(\frac{\pi}{8}\right)\right] \quad (10.9c)$$

where $erfc$, the complementary error function, is as already defined (Equation 10.6b).

The theoretical and measured variation pattern of $P_e I_{8\phi}$ as a function of E_b/N_o show that around $\frac{E_b}{N_o} = 10 \, dB$, a gain of 3-orders of magnitude (from 10^{-3} to 10^{-6}) is achieved by coding for 8-phase PSK.

The fact that 8ϕPSK implementation provides high bandwidth efficiency, makes it the most common choice for satellite communications, typically yielding a BER value of 10^{-10}. A coding gain of about 2 dB is achievable under 8ϕPSK.

Recent INTELSAT (and other) satellite communications systems typically use transmission of Mbps-range for information bits, achieving a BER of 10^{-10} (comparable to optical fiber systems) using a transponder bandwidth of a few MHz.

10.4. Digital Hierarchy

A particular country or region usually adopts for its digital voice network different types of hierarchies in terms of different sets of combinations of numbers of multiplexed voice channels and corresponding bit rates, to be used as standards.

For instance, the digital hierarchies and their standard designations employed in North America and Europe are as shown in Table 10.3.

Table 10.3. Digital Hierarchies of North America and Europe.

North America				Europe			
Number of Voice Channels	Bit Rate (Mbps)	Designation of Level	Equivalence	Number of Voice Channels	Bit Rate (Mbps)	Designation of Level	Equivalence
24	1.544	DS1	DS1	30	2.048	E1	E1
48	3.152	DS1C	2xDS1	120	8.448	E2	4xE1
96	6.312	DS2	4xDS1	480	34.368	E3	4xE2
672	44.736	DS3	7xDS2	1920	139.264	E4	4xE3
4032	274.176	DS4	6xDS3	7680	565.148	E5	4xE4

10.5. Multiple Access and Related Problems and Performance

10.5.1. Introduction

A communication satellite receives the signal transmitted by a multiple set of earth stations ('uplink') and, after amplifying it by many orders of magnitude, transmits it to a set of multiple receiving earth stations ('downlink'). The amplification process in the satellite is performed by a set of 'repeaters' such as the traveling wave tubes (TWTs), coupled-cavity TWT (CC/TWT), solid state power amplifier (SSPA), or some similar device. The properly amplified signal is then transmitted using downlink satellite antenna(s) to the destination earth stations.

The technique of how a number of earth stations can utilize the satellite frequency band is the subject of this Section. These techniques include the frequency division multiple access (FDMA), time division multiple access (TDMA), satellite switched TDMA (SS/TDMA), code division multiple access (CDMA), single-channel-per-carrier

(SCPC), and demand assigned (DA), among others. These multiple access schemes and related performance are discussed below.

10.5.2. Frequency Division Multiple Access (FDMA): Intermodulation

At a particular time, a number of transmitting earth stations may be simultaneously present in the satellite repeater, both for uplink as well as for downlink. TWT is a wideband device and can therefore have its total frequency bandwidth shared by multiple earth stations. Each of the transmit earth stations bundles its signals destined for various receiving earth stations, and modulating a single carrier with them – a process referred to as frequency division multiplexing (FDM). The modulation is also performed in the frequency domain, and is therefore called frequency modulation (FM). The carriers of various transmitting earth stations are 'permanently' assigned a particular frequency (or a small range of frequency) to allow FM, with small frequency 'guard bands' separating various carriers to minimize interference between carriers. The guard bands are also of course in frequency domain. This frequency-modulated, frequency division multiplexed, frequency division multiple access (FM/FDM/FDMA) signal and carrier pattern allows FDMA in the uplink.

The satellite de-multiplexes the uplink signal in order to bundle the signal (carriers) to the same downlink (destination) earth station, transmitting the entire set of downlink carriers to various receiving earth stations. The complete configuration of the carriers received (uplink) and transmitted (downlink) remains unchanged in time. This is schematically represented in Figure (10.2).

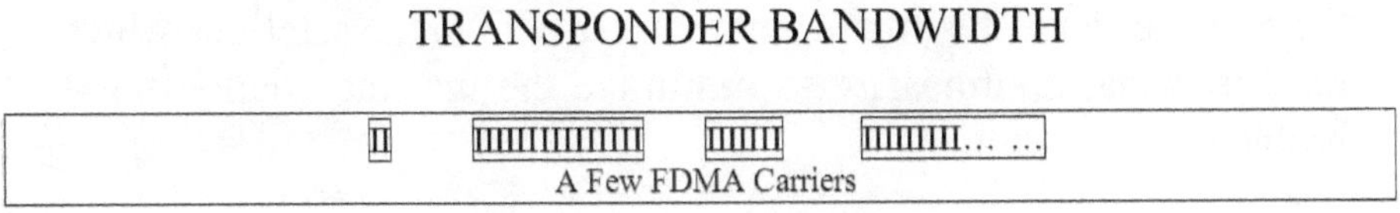

TRANSPONDER BANDWIDTH

A Few FDMA Carriers

Figure 10.2. The general concept of FDMA.

The FM/FDM/FDMA technique applied for various types of communications services (voice, television, data) was widely used until advent of digital communications during the last several decades. This (FDMA) multiple access method suffered from several problems in terms of efficiency of the use of the overall available frequency bandwidth – a limited natural resource. A particular source of inefficiency and source of interference noise is the intermodulation process that is inherently present in a non-linear device such as TWT, and disallows the full use of the satellite power, as explained below.

Intermodulation

Intermodulation noise interferes with the actual signal and represents inefficiency in the TWT power utilization. As explained earlier, this type of noise, associated with generation of intermodulation products (IMP), results under simultaneous presence of two specific conditions:

(1) Two or more carriers are present in the amplifier (satellite TWT) at the same time,
(2) The amplifier has a nonlinear characteristic in its output performance.

Multiple carriers in a nonlinear device may be employed in the case of digital carriers as well, which will then suffer disadvantage (IMP) as in the case of analog carriers. The mechanism of generation of the second order IMP has been explained in this book earlier. Higher (third and higher) order non-linearity produce more complex set of IMP, even though their strength may be progressively lower.

Some of the IMPs can be eliminated from the output with the use of proper bandpass filter(s), but this represents extra weight for the satellite. Thus it is important to use TWTs in the satellites which have minimal nonlinearity to minimize the intermodulation noise problem.

Nonlinearity is more pronounced in the high power range. Since it is usually not possible to avoid nonlinearity altogether, a TWT is commonly operated at a power-level which is smaller than its saturation (maximum) power level. Reducing the operating power level as compared to the maximum power level of the amplifying device is referred to as the back-off. With adequate amount of back-off, the operating point lies in the linear region, and the intermodulation noise is no longer a serious problem. However, back-off implies that the available TWT power becomes reduced, representing system inefficiency.

The power inefficiency and the intermodulation noise problem is one of the biggest drawbacks of the FDMA system.

10.5.3. Time Division Multiple Access (TDMA)

An obvious remedy of the intermodulation noise problem present in an FDMA system is to avoid the simultaneous use of two or more carriers in the transponder. This means that at any given instant of time, only a single carrier is present in the transponder. Different carriers then use the transponder in turn, sharing its full capacity in the time domain. For system efficiency, each user (transmitting earth station) uses the transponder for a very small interval of time. The actual length of time allocated to different users from a set sharing the TWT capacity depends on the volume of its 'traffic' requirement. To fully utilize the maximum available capacity of the transponder, therefore, each user transmits its traffic at a very high bit rate or speed. This is possible only in digital transmission form.

Digital transmission by a participating earth station at a very high bit-rate consistent with the maximum available power level of the TWT (i.e., at saturation, without the need of back-off, since no intermodulation within the pass-band occurs when only a single carrier is present in the transponder at any given instant of time), allows efficient utilization of maximum available power level. Time-sharing of the transponder by a set of multiple users, using a single carrier in its turn, at a very high bit-rate (typically in the range of Mbps, or 10^6 bit/sec), is referred to as time *division multiple access (TDMA)*, and the high bit-rate, ultra-short-duration (~μsec) transmission from a particular user as its TDMA burst. TDMA can be used for fixed satellite services (FSS) as well as for mobile satellite services (MSS). TDMA equipment are usually quite expensive, and hence not suitable for hand-held terminals providing mobile telecommunications services. For the satellite and terrestrial mobile services, usually code division multiple access technique is employed.

A judicious distribution of TDMA burst within the TDMA frame in the time-domain (analogous to the arrangement of multiple carriers simultaneously in the frequency domain in the case of FDMA) is essential to maximize the systems efficiency. One drawback of a system with digital modulation, digital multiplexing, and TDMA burst mode operation in the time domain, e.g., QPSK/TDM/TDMA, (analogous to FM/FDM/FDMA in the frequency domain) is that now the earth stations have to use high-cost TDMA equipment for high bit-rate digital transmission and reception. However, TDMA is still used by many service providers (including INTELSAT) for suitable applications. Algorithms have been developed for efficient use of the satellite power in the TDMA mode[47] which are employed by INTELSAT in its TDMA operation. A variation of TDMA mode, referred to as satellite-switch TDMA (SS/TDMA) where the beam connectivity of a transponder is periodically reconfigured to different beam regions within the TDMA frame to serve users within disjoint beam regions, is also useable. In any case the set of bursts that are periodically used in a transponder define its *TDMA frame*.

Analogous to the FDMA guard-bands, guard-times are used in the time domain in the case of TDMA. Also, for proper identification of the transmission (uplink) and reception (downlink) and other necessary functionalities, extra segments of small duration called preambles, are transmitted (and received) in front of each TDMA burst.

Under the use of time-guards and preambles within the TDMA frame and of similar overhead (e.g., reference bursts for synchronization of all users in a TDMA frame), the *TDMA frame efficiency*, a figure of merit of the TDMA system for a network of users, is readily evaluated in terms of the ratio of the actual traffic bits to the total number (including overhead) of bits transmitted by a transponder within a TDMA frame. Thus, using the following notations for the parameters involved:

F = TDMA frame-length;

B = Bit-rate of the TDMA system (in bits/sec);

P = number of bits in a preamble;

N = number of bursts/number of preambles (= number of users);

G = number of bits used for each guard-time;

N_g = number of guard times within a TDMA frame;

N_r = number of reference earth stations;

B_r = number of bits used in a single reference burst;

the total overhead, in terms of the total number of (non-traffic) bits (B_x) within a single TDMA frame can be simply written as

$$B_x = N_r B_r + NP + N_g G + N_r G$$

The first term on the right is the total number of bits for reference functionality by N_r reference earth stations; the second term is the total number of bits for preambles; the third term is the number of bits for all the guard-times within the frame; and the fourth term is the net number of bits for guard-times associated with the reference bursts.

The overall number of bits transmitted within a TDMA frame is, of course, simply $B_T = BF$; so that the difference $(B_T - B_x)$ represents the number of bits used within the frame for actual traffic. Consequently, the TDMA frame efficiency ϵ, can be simply written as the ratio

$$\epsilon = \frac{B_T - B_x}{B_T}$$

A TDMA frame length of several milliseconds with about 20 users or so can achieve high efficiency (~95%). Small values of the TDMA frame-length are likely to yield low efficiency. As the users transmit (and receive) the burst data only during short duration in their turn, they remain inactive for a substantial portion of the frame length, keeping their incoming (or outgoing) data in memory. Long frames offer higher efficiency necessitate high-capacity memory for the earth stations corresponding to a longer wait for their turn within the frame. High-capacity memory devices have become more easily and economically available with advances in the microchip technology during recent times, and such advances continue with the passage of time. Thus, TDMA is likely to become even more prevalent and efficient than it is today. With the trend of continuing expansion of digital communications, the use of TDMA transmission for various suitable applications is also expected to grow.

10.5.4. Code Division Multiple Access (CDMA) in Mobile Channels

10.5.4.1. General

Code Division Multiple Access (CDMA) mitigates the impact of channel interference through the *spread-spectrum* modulation technique.

After the Standard for TDMA in North America (NA-TDMA system or IS-54) was established, the development in the CDMA technology was started in 1989. The feasibility of using CDMA in the mobile/cellular system was demonstrated in November 1989, and a standard was issued on December 9, 1992, viz.

IS-95, "Mobile Station-Base Station Compatibility Standard for Dual Mode Wideband Spread-Spectrum Cellular System."

Some of the basic features of the above system (IS-95) are specified below. The following definitions of terms and specifications are useful in this connection.

System Time – The CDMA system reference time is synchronized to the Universal Time Coordination (UTC), using the same time origin as the GPS (Global Positioning System) time. All Base Stations (BSs) use the same system time; while all Mobile Stations (MSs) use this same system time, offset by the propagation delay from the BS to the MS.

CDMA Channel Number – This is an 11-bit number corresponding to the center of the CDMA frequency assignment.

Code Channel – This refers to a sub-channel of a forward CDMA channel containing 64 code bits.

Code Symbol – This refers to the output of an error-correcting encoder.

Forward CDMA Channel – Refers to one or more code channels.

Active Set – This refers to the set of pilot channels associated with the CDMA channels containing forward traffic channels assigned to a particular mobile station (MS).

Frame offset – This is the time offset of the traffic channel frames from system time in integral multiple of 1.25ms (the maximum frame offset is 18.75ms).

Handoff (HO) – This refers to transfer of communication with the mobile station (MS) from one base station (BS) to another base station. The HO are of three types: Hard HO, Soft HO, and Idle HO.

Hard HO occurs under one or more of the following conditions:

(1) The MS is transferred between disjoint active sets;
(2) The CDMA frequency assignment changes;
(3) The frame offset changes;
(4) The MS is transferred from a CDMA traffic channel to an analog voice channel (not vice versa).

Soft HO occurs when the CDMA cell changes without a change in the CDMA frequency.

Idle HO occurs when the paging channel is transferred from one base station (BS) to another BS.

Long Code – This refers to a pseudo-noise (PN) sequence with period $(2^{42} - 1)$ using a tapped n-bit shift-register.

Layer – The communication protocol with reference to a peer layer is organized using a layer structure.
Layer 1 – This refers to the physical layer that presents a frame by the multiplex sub-layer and transforms it into an over-the-air waveform.
Layer 2 – Provides for the correct transmission and reception of signaling messages.
Layer 3 – This provides the control of the cellular telephone system through signaling messages (which originate and terminate at this layer).

10.5.4.2. CDMA Output Power

The power levels in a CDMA system must be properly controlled for satisfactory operation.

Mobile Station (MS) – The *mean output power* of the MS is specified to be less than $-50\ dBM/1.23\ MHz$ ($-111\ dBM/Hz$) for all frequencies within $\pm 615\ kHz$ of the center frequency.

The <u>Gated Output Power</u> of the MS is typically within certain limits (envelope); for example, with the transmitter noise floor less than $-60\ dBM/1.23\ MHz\ (-121\ dBM/Hz)$.

10.5.4.3. Modulation of CDMA Traffic Signal

Since the mobile station (MS) has no established system time as at the base station (BS), the modulation characteristics are different for the forward channel and the reverse channel.

The Data rates of the access channel and the reverse channel, along with the modulation and other parameters, are listed in Table 10.4. Note that the access channel has only one data-rate (4800 sps) but the reverse channel has four data-rates, viz., 1200, 2400, 4800, and 9600 symbols/sec (sps).

Table 10.4
Access Channel and Reverse Channel Modulation Parameters

	Access Channel	Reverse Channel
Modulation	6 code symbol/mod. symbol	
Modulation Symbol Rate	4800 sps	1200, 2400, 4800, 9600 sps
Code Symbol Rate	28800 sps	
Code Rate	1/3 bits/code symbol	
Modulation Symbol Duration	208.33 μs	
Walsh Chip Rate	307.20 kcps	
PN Chips/Code Symbol	42.67	
PN Chips/Modulation Symbol	256	
PN Chips/Walsh Chip	4	
PN Chip Rate	1.2288 Mcps	
Data Rate	4800	9600 4800 2400 1200 bps
Transmit Duty Cycle	100	100 50 25 11.5 %

10.5.4.4. Transmission Sequence

At 9600 bps, the transmission sequence is to send row-by-row, up to row 32, in the following sequence:

    1 2 3 4 ... 30 31 32.

At 4800 bps, the following unique order of rows is sent:

    1 3 2 4   5 7 6 8   9 11 10 12   13 15 14 16   17 19 18 20
    21 23 22 24   25 27 26 28   29 31 30 32

Each of the above sequences can be represented symbolically as follows:

    J, J+2, J+1, J+3

with

$$J = 4i + 1, \ i = 0, 1, 2, 3, \ldots (32/4 - 1)$$

At 2400 bps, the transmission sequence follows the unique order of rows:

    J, J+4, J+1, J+5, J+2, J+6, J+3, J+7

with

$$J = 8i + 1, \ i = 0, 1, 2, 3, \ldots (32/8 - 1).$$

At 1200 bps, the unique sequence of rows is:

J, J+8, J+1, J+9, J+2, J+10, J+3, J+11, J+4, J+12, J+5, J+13, J+6, J+14, J+7, J+15

with

$$J = 16i + 1, \qquad i = 1, 2$$

For Access Channel (4800 bps), the following unique order is followed by the interleaver:

J, J+16, J+8, J+24, J+4, J+20, J+12, J+28, J+2, J+18, J+10, J+26, J+6, J+22, J+14, J+30

with

$J = 1, 2.$

The <u>Interleaving Algorithm</u> is as shown below in Table 10.5.

Table 10.5.
Interleaving Algorithm (Adapted from William Y.S. Lee, *ibid*, p.154, by Permission from McGraw Hill)

Row/Column	1	2	3	4	5	6	7	8	9	10	11	12	13	14	15	16	17	18
1	1	33	65	97	129	161	193	225	257	289	321	353	385	417	449	481	513	545
2	2																	
3	3																	
4	4																	
5	5																	
6	6																	
32	32	64	96	128	160	192	224	256	288	320	352	384	416	448	480	512	544	576

<u>Walsh Code</u> – The reverse channels have orthogonal codes. The 64-ary Walsh codes employed consists of 64 codes each 64 bits long. Every sixth symbol interpreting the Walsh code of 64 chips is sent out.

12 bits = 36(=12x3) code symbols
= 36/6 modulation symbols
= 1 power control group

The access channels and the reverse channel are direct-sequence spread by the long code prior to transmission. The long code is periodic with period $2^{42} - 1$ chips satisfying the linear recursion specified by the polynomial $p(x)$ given as:

$$p(x) = x^{42} + x^{35} + x^{33} + x^{31} + x^{27} + x^{25} + x^{22} + x^{21} + x^{19} + x^{18}$$
$$+ x^{17} + x^{16} + x^{10} + x^7 + x^6 + x^5 + x^3 + x^2 + x^1 + 1$$

Each PN chip of the long code is generated by a 42-shift register long code generator.

<u>Quadrature Spreading</u> sequences are periodic with a period of 2^{15} chips, and the spread polynomials of channel 1 and Q pilot PN sequences are as follows:

$$p_1(x) = x^{15} + x^{13} + x^9 + x^8 + x^7 + x^6 + x^5 + 1$$

$$p_Q(x) = x^{15} + x^{12} + x^{11} + x^{10} + x^6 + x^5 + x^4 + x^3 + 1$$

These are of period $2^{15} - 1$. The PN sequence repeats every 26.66 ms ($2^{15}/1228800$ sec). In 2 seconds, the number of repetitions is 75. Reverse CDMA channel 1 and Q mapping for an offset QPSK modulation is also used.

<u>10.5.4.5. Call Processing</u>

The MS call processing consists of the following states:

- The MS selects which system to use
- It acquires the pilot channel of a CDMA system within 20 ms.

- It obtains system configuration and timing information for a CDMA system.
- It synchronizes its timing to that of a CDMA system.

10.6. Digital Television

10.6.1. General

Consideration of digital transmission of video and television programming was initiated some 40 years ago, but was accelerated only recently. The digital TV satellites —ECHOSTARs — were launched in late 1990s with the first ECHOSTAR transmission systems parameters (and the parameters of the small earth-terminals—usually consisting of about 45 centimeter diameter 'roof-top' antenna and receiver system) specified by the author of this book (A.K.Sinha) around 1996. ECHOSTAR provides digital TV service under the commercial brand name 'DISH Network.' A host of other satellite and terrestrial (fiber-optical cable) systems and the service providers at present include DirecTV (Hughes satellite), Atlantic Telephone and Telegraph Company (AT&T), and COMCAST. *High Definition television (HDTV)* is now offered by all the providers as a popular option, for improved picture quality, especially for video involving fine details and rapid motion (sports-related broadcast). Internet services including U-Tube and mobile video have also become very common and popular. News channels such as CNN using satellite news gathering (SNG) techniques employing van-and-truck-mounted earth terminals routinely make any selected news event broadcast available to the whole world instantaneously. In general, a variety of *INTERNET-based* service applications via satellites, involving video, audio, teleconferencing, and texts, have truly transformed the world into a *'Global Village,'* with new technical innovations and advances in service applications occurring virtually daily.

Special techniques, particularly pertaining to bit compression, motion-compensation, and color signal processing, for example, in combination with HDTV with twice the sampling rate (and hence double the bit rate) compared to conventional TV, have succeeded in making TV signal life-like. Explorations into the possibility of three-dimensional (3-D) video, is already continuing for motion pictures in cinema halls, and for residential television viewers. It may be mentioned in passing that frequency-modulated (analog) television (TV/FM), the only mode of TV transmission and broadcast available just a few decades ago, and considered the only possible mode for this purpose by many leading engineers (*"Digital TV will never work"*) is now nearly obsolete (except for some users who use FM-to-digital converter devices). The advent of digital TV has made it possible to multiply the number of TV channels into many hundreds in place of just a few TV/FM channels before. Combination of video, voice, news captions, streaming information, and commercial advertisements can be seen on a variety of TV screen designs.

10.6.2. Typical Video Parameters

While technological innovations continue to enhance systems efficiency and diversity of applications regularly, a set of systems parameters shown in Table 10.6 can be taken as basic characteristic combinations (based on 1990s technology.)

Table 10.6[48]
Basic Parameters of Digital Video and Television

Bit Rate	Modulation	Bandwidth	E_b/N_o $(dBHz)$	BER
56-1536 kbps	QPSK	31-60 kHz	11.6	
20 Mbps	QPSK	11.2 MHz	13.6	
45 Mbps	QPSK	23.2 MHz	13.6	
130 Mbps	Coded 8ϕ	72 MHz	11.5	

10.6.3. Video Color

As is well-known, the television screen is basically composed of a large number of picture elements ("pixels") which are scanned as TV program recording is made. These pixels are scanned in the same sequence to regenerate the recorded picture.

Each pixel has a characteristic color when recording and the display for the viewer take place. The color is composed of a mixture of three basic colors: Red (R), Green (G), and Blue (B). The amounts of the three component (basic) colors for a particular pixel in the course of a particular recording/display of the picture at high speed determine the actual resultant color or hue of the pixel, and thus the entire set of pixels within the screen yields a specific picture with color where the color might vary from one pixel to the next. Thus, the (R,G,B) combination for each pixel provides the full-screen pictures from frame-to-frame, with vivid colors. Digital TV, benefitting from the signal integrity associated with bit-regeneration process is capable of a wider range of colors with a high level of fidelity, compared with the old (analog) method (FDMA/FM/TV).

10.6.4. High Definition Television (HDTV)

High Definition TV (HDTV) simply refers to the case where the scanning rate of twice the usual TV is employed, by allowing the number of horizontal lines scanned at recording and display at twice the usual or conventional method. This permits a greater resolution and hence an improved quality of the TV picture. Now-a-days, commonly the HDTV is the preferred mode of provision of programming to viewers. This is especially important for fine details and/or fast motion, such as in the case of recording and display of sports events.

References:

[44] H. Taub and D.L. Schilling, Principles of Communication Systems, McGraw Hill, New York, 1971.
[45] Pritchard et al, *ibid*, p. 347
[46] W.C.Y. Lee, Mobile Communications Design Fundamentals, Howard W. Sams, Indianapolis, Minnesota, 1986.
[47] A.K. Sinha, "A Model for Burst Assignment and Scheduling," COMSAT Technical Review, Vol 6, No. 2, P. 219, Fall 1976.
[48] Pritchard et al, *ibid*, p. 354.

11. SPECIAL TOPICS IN MOBILE SERVICE SYSTEMS

11.1. Satellite Link Design

The basic criterion for the transmission link design in satellite communications systems is the carrier-to-noise ratio (C/N) or, equivalently, the energy-per-bit-to-the-noise-density ratio (E_b/N_o) in the case of digital satellite systems. It may be recapitulated that the two are simply related, since

$$\frac{C}{N} = \frac{E_b \times R_b}{N_o \times B} = \left(\frac{E_b}{N_o}\right)\left(\frac{R_b}{B}\right) \quad (11.1)$$

where R_b is the bit rate, and B is the bandwidth.

If the total noise is the resultant of the contributions due to a number of independent noise sources, it is easy to obtain the net carrier-to-noise, $(C/N)_T$, by first summing up the various noise contributions. Thus, the total noise, N_T, can be simply written as

$$N_T = \sum_i N_i$$

where N_i is the ith noise contribution.

In a satellite link, noise is contributed mainly due to the following factors:

(1) Noise (N_u) is due to the uplink (i.e., in the transmission link from the transmitting earth station to the satellite, measured at the satellite.)
(2) Noise (N_I) is due to intermodulation products when in a nonlinear satellite repeater, sets of multiple carriers are present.
(3) Noise (N_i) due to intersystem interference, arising when the neighboring satellites in the orbit (geostationary or LEO satellite systems) are close enough to interfere with one another.
(4) Noise (N_d) in the downlink (i.e., in the transmission from the satellite to the receiving earth station, measured at the receive earth station.)

The uplink and downlink noise generally include the thermal noise, and weather (snow, rain, etc.)-induced absorption losses and noise. Similarly, the interference noise (N_i) itself may be a sum of contributors due to a multiplicity of neighboring satellites, and intentional or inadvertent jamming effect from other transmitting earth stations. Here, it suffices to write N_T as the four contributions described above.

$$N_T = N_u + N_d + N_I + N_i \qquad (11.2b)$$

Generalization of Equation (11.2b) to include the cases when one or more of the four factors on the right of Equation (11.2b) is the sum of two or more constituent contributions is obvious.

Now, the carrier-to-noise ratio could be written as

$$\frac{C}{N_T} = \frac{C}{N_u + N_d + N_I + N_i}$$

i.e.,

$$\frac{N_T}{C} = \frac{N_u + N_d + N_I + N_i}{C}$$

$$= \frac{N_u}{C} + \frac{N_d}{C} + \frac{N_I}{C} + \frac{N_i}{C}$$

Now, using the inverse of the ratios with a power of (-1), i.e., writing

$$\frac{N_T}{C} = \left(\frac{C}{N_T}\right)^{-1}$$

with similar inverse representations of the ratios on the right, we can write:

$$\left(\frac{C}{N}\right)_T^{-1} = \left(\frac{C}{N}\right)_u^{-1} + \left(\frac{C}{N}\right)_d^{-1} + \left(\frac{C}{N}\right)_I^{-1} + \left(\frac{C}{N}\right)_i^{-1}$$

(11.3a)

As mentioned, generalization of Equation (11.3a) when additional noise contributions are present, or when any of the four noise contributions is itself an expression of a combination of multiple sub-contributions, is self-evident.

The Equation (11.3a) can of course be written in terms of the (E_b/N_o). Thus, when each (C/N) term on the both sides of the equation is replaced by its corresponding (E_b/N_o) value, we have

$$\left(\frac{E_b}{N_o}\right)_T^{-1} = \left(\frac{E_b}{N_o}\right)_u^{-1} + \left(\frac{E_b}{N_o}\right)_d^{-1} + \left(\frac{E_b}{N_o}\right)_I^{-1} + \left(\frac{E_b}{N_o}\right)_i^{-1}$$

(11.3b)

Equations of the above type are very useful to obtain the resultant $(C/N)_T$ or the resultant $(E_b/N_o)_T$. Also, it is clear that this pattern of including different noise sources applies to fixed satellite services (FSS) as well as to mobile satellite services (MSS); and analogous results hold for satellite networks as well as terrestrial or hybrid networks.

11.2. Satellite Antenna Beam Coverage Pattern

For FSS or MSS, the satellite (downlink beam) antenna of a geostationary satellite generates a beam coverage which must be designed to properly envelope the desired service region. This can be done in a number of ways. For geometrically regular or well-defined (service area or) beam shapes, such as circular or elliptical, the design of the satellite downlink antenna is relatively simple. For such geometrically regular coverage regions, as well as for arbitrary and geometrically irregular coverage region, a most general alternative is to define a set of points at the periphery of the coverage contour on the surface of the earth. Here, we provide the method of determining the positions of any number, n, of points distributed over the contour of the beam coverage region.[49]

For a beam coverage region of practical interest, any analysis of the coverage pattern—say, the determination of the latitudes and longitudes of a reasonable number of points around the contour, for example—becomes rather complex mathematically since the earth's spherical shape must be taken into account, and the spherical geometry must be employed in such an analysis, as illustrated below.

Here we use the following notations:

B_o = the satellite antenna bore-sight (the central point of the coverage)

S_o = the sub-satellite point (lying on the equator for a geostationary satellite))

N = total number of selected points on the contour

n = serial number of a point on the contour ($n = 1, 2, 3, ... N$)

P_n = designation of the nth point

B_n = angular beamwidth of half-power (or of any specified power-level) in the direction of the nth point. Note that for a circular beam, $B_n = B_\varrho$, where B_ϱ is a constant; i.e., $B_1 = B_2 = B_3 = \cdots = B_N = B_\varrho$.

δ_b = tilt angle of the satellite antenna bore-sight (B_o)

δ_n = tilt angle of P_n

ϕ_o = longitude of P_n

θ_b = latitude of the satellite antenna bore-sight

ϕ_b = differential longitude of B_o ($\phi_b > 0$ if the bore-sight lies west of S_o; $\phi_b = 0$ if the longitude of B_o equals the longitude of S_o)

α_n = the angle between the lines $B_o P_n$ and $B_o S_o$ (measured in the plane containing B_o, S_o, and P_n)

θ_n = latitude of P_n

ϕ_n^1 = longitude of P_n relative to $S_o = \phi_n - \phi_o$

i.e., $\phi_n = \phi_o + \phi_n^1$ = actual longitude of P_n

Then it can be shown[49] that the following relations hold for the bore-sight B_o:

$$\delta_b = \tan^{-1}\left[\frac{\sin \beta}{q - \cos \beta}\right] \qquad (11.4a)$$

where

$$q = 6.6235 \qquad (11.4b)$$

and

$$\cos \beta = \cos \theta_b \cos \phi_b \qquad (11.4c)$$

while for the nth point P_n on the beam coverage contour, the following relations hold:

$$\cos \delta_n = \cos \delta_b \cos B_n + \sin \delta_b \sin B_n \cos \alpha_n$$

$$(11.5a)$$

Introducing now three angles $U_n, W_n,$ and V_n, and as follows:

$$U_n = \sin^{-1}(q \sin \delta_n) - \delta_n \qquad (11.5b)$$

$$W_n = \cot^{-1}(\sin \delta_b \cot B_n - \cos \delta_n \cot \alpha_n)$$

$$(11.5.c)$$

$$V_n = W_n + \tan^{-1}(\sin \phi_b \cot \theta_b) \qquad (11.5.d)$$

The latitude and longitude of the point P_n are given as

$$\theta_n = \sin^{-1}(\sin U_n \cos V_n) \qquad (11.6a)$$

$$\phi_n = \tan^{-1}(\tan U_n \sin V_n) \qquad (11.6b)$$

The total number of roughly evenly distributed points, N, on the contour can be selected to provide adequate information about the shape of the beam coverage pattern, with proper resolution. Different points $P_1, P_2, ..., P_N$ correspond to successive values of the angles α_n in the full range $(0° < \alpha_n < 360°, n = 1, 2, ..., N)$. For each selected point P_n, with the selected or specified value of α_n ($1 \le n \le N$), the above analysis is repeated to get the latitude and longitude of P_n (θ_n, ϕ_n), and the complete contour generated. In case of multiple beams, the above analysis is carried out for each beam. In particular, for an elliptical beam with maximum beam half-width B (corresponding to the semi-major axis of the ellipse), and the minimum half beam-width B^1 (corresponding to the semi-minor axis), the following relation holds for the half beam-width corresponding to any point P_n on the elliptical contour:

$$\cot^2(B_n) = \cot^2 B_1 \cos^2(\alpha + \alpha_n) + \cot^2 B_2 \sin^2(\alpha + \alpha_n)$$

where α is the angle between the major axis of ellipse and the direction of the azimuth of the bore-sight B_o.

For a circular beam contour, since $B = B_1 = B_2$, Equation (11.7a) above reduces to the identity, as expected

$$\cot^2 B_n = \cot^2 B \qquad (11.7b)$$

i.e.,

$$B_n = B, n = 1, 2, \ldots, N$$

where B is the constant value of the half power beam width for any and every point P_n on the circular contour.

11.3. Cellular Pattern of Multiple Spot Beams

A case of special interest, particularly for an overall beam pattern composed of a fairly large number of closely packed spot beams generated by a geostationary satellite, is when these component spot beams are of hexagonal shapes, like a beehive. Thus, each spot beam is surrounded by six similar spot beams, except for the edge of the beams. Clearly equal size hexagonal figures can fill a two-dimensional area, without leaving any region within the overall boundary. Such honeycomb pattern is therefore best suited for a multiple spot beam generated by a multi-feed satellite antenna, each feed generating a spot beam, to provide a contiguous set, as illustrated in Figure 11.1, for a partial subset of the honeycomb pattern of multi-spot beams, numbers A1, A2, A3, . . . and B1. B2, B3, . . . where A and B designate two different satellite systems (two different providers, for example), and the honey-comb structures of the cellular coverage regions are underlaid and overlaid, as illustrated here. Other types of arrangements are also possible.

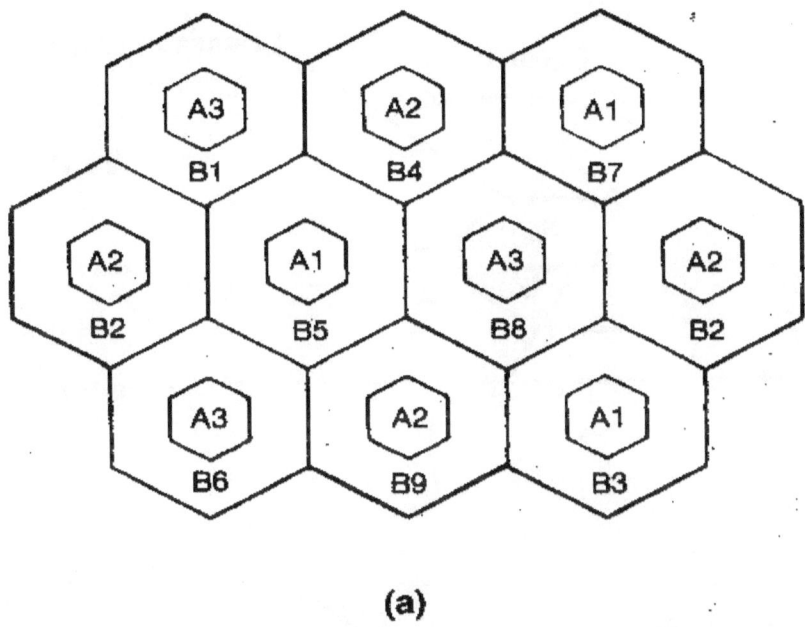

(a)

Figure 11.1. A honeycomb pattern of multi-spot beams
(Adapted from W.Y.S. Lee, *ibid*, p. 516, by Permission from McGraw Hill).

It should be noted that the above type of multi-spot beam patterns are also commonly employed for provision of cellular/mobile communications services, with the help of low earth orbit (LEO) satellites. A multiple set of satellites with carefully chosen orbital parameters, and orbited such that, as one satellite moves out of range with respect to the coverage zone, the next satellite in the set sequentially assumes an orbital position in its passage to continue the coverage of the cellular beam region, without any gap. Thus, usually a continuous coverage by successive satellites in the series can be assumed to be available, in order to provide the mobile or cellular communications services.

In order to express the relevant mathematical relations[50-51], consider the planar representation of the multi-spot beams as shown in Figure 11.2.

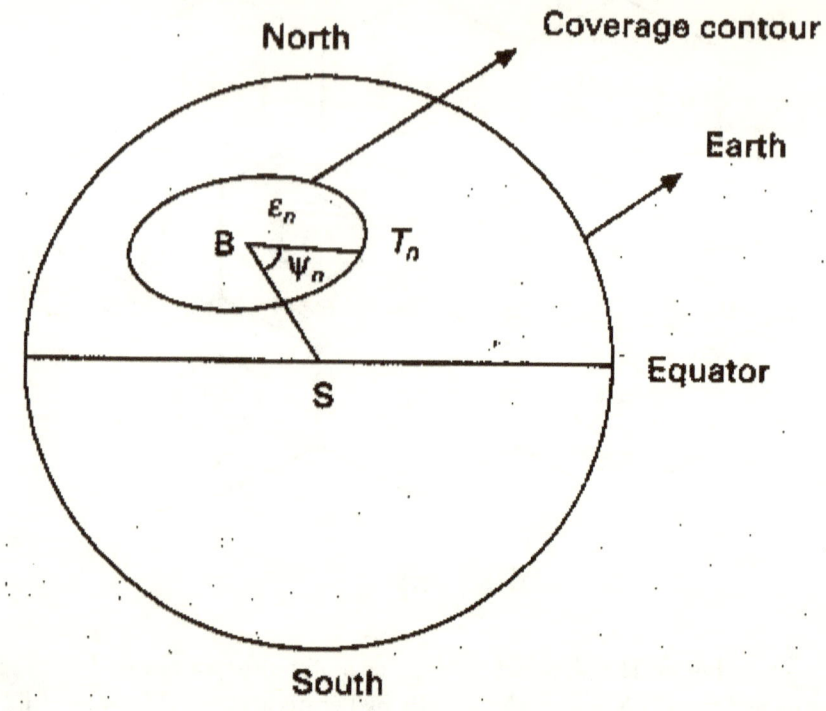

Figure 11.2. The multiple spot beam configuration (Adapted from M. Richharia, *ibid*, P.434, by Permission from McGraw Hill).

Actually, Figure 11.2 is the equatorial planar intersection of the earth showing the net angle (ξ_o) subtended by the coverage region at the center (O) of the Earth. The angles subtended by various spot beams at the satellite location (S), i.e., the half-power beamwidths of various spot beams, labeled $\beta_1, \beta_2, \ldots$, are also indicated. Also, let (see Figure 11.2):

R_o = Earth's radius

E_o = minimum elevation angle (corresponding to one of the outer-most spot beam)

δ = geocentric angle of each spot beam

H = altitude of the satellite

β_i = beamwidth of the ith spot beam ($i = 1, 2, 3, ...$)

$\xi_o = \angle POQ$ = angle subtended by the composite beam at the center of the earth (O)

N = total number of cells

θ_o = beamwidth of the central cell

θ_n = beamwidth of the nth 'crown'

For a symmetrical composite beam pattern with respect to the line OS, noting that

$$\angle PSO = \sum_i{}^{N/2} \beta_i = B/2$$

where B is the overall beamwidth, and the prime over the summation sign denotes summing of the angles β_i only on one side of the line OS, and also that

$$\angle SPO = \frac{\pi}{2} + E_o$$

and finally

$$\angle SOP = \xi_o/2$$

we have from ΔSPO, the simple relationship

$$\frac{B}{2} + \left(\frac{\pi}{2} + E_o\right) + \xi_o/2 = 2\pi$$

i.e.,

$$B + 2E_o + \xi_o = 3\pi \qquad (11.8a)$$

Also, it can be easily shown that

$$\xi_o = 2R_o\left[\frac{\pi}{2} - E_o - \sin^{-1}\left(\frac{R_o}{R_o+H}\cos E_o\right)\right]$$

$$(11.8b)$$

It can also be shown from the geometry of Figure (11.3) that for low enough values of the satellite altitude ($H \ll R_o$) and the minimum elevation angle E_o, the composite beamwidth can be simply approximated as

$$B \simeq 3\pi + 2R_o(E_o + \pi) \qquad (11.8c)$$

Additional useful relations in case of such a symmetrical multi-spot beam antenna coverage configuration consisting of a total number, N, of the cells are given below:

$$N = 1 + 3n(n+1) \qquad (11.9a)$$

$$\delta = 2\xi_o/(2n+1) \qquad (11.9b)$$

$$\theta_o = 2\tan^{-1}\left[\frac{\sin\frac{\delta}{2}}{1-\cos\frac{\delta}{2}+\frac{h}{R}}\right] \qquad (11.9c)$$

and

$$\theta_n = \tan^{-1}\left[\frac{\sin \xi}{1-\cos \xi + \frac{h}{R}}\right] - \frac{\theta_o}{2} - \sum_{k=1}^{n-1} \theta_k$$

(11.9d)

where 'crown' refers to the number n, determining the total number, N, of the hexagonal cells.

11.4. Doppler Effect and Range-Rate

A geostationary satellite (GEO) remains fixed with respect to the earth, so there being no relative motion between the earth and such satellites (except for temporary drift motion, if any), there is no Doppler shift for geostationary satellites. However, low-earth orbit (LEO) and medium-earth orbit (MEO) satellites, with significant amounts of relative velocities with respect to the earth, suffer from Doppler Effect, and necessary corrections to the uplink and downlink frequencies must be made as part of their operations. The case of a high earth orbit (HEO), where the satellite altitude may be higher than the geostationary altitude, may also be treated similarly.

Obviously, the frequency shift (increase in the frequency value as the satellite is approaching a particular earth station; decrease as it is receding with respect to the earth station), Δf, due to the Doppler effect can be generally written as

$$\Delta f = \pm \frac{v}{c} f_o$$

(11.10a)

where v is the relative velocity involved, c is the velocity of light, and f_o is the original frequency value.

Also, clearly, the relative velocity is approximately given as

$$v \cong \frac{R(t_2)-R(t_1)}{t_2-t_1} \qquad (11.10b)$$

where t_1 and t_2 are two time-instants ($t_2 > t_1$; $t_2 - t_1$ arbitrarily small), and $R(t_1)$ and $R(t_2)$ are the corresponding ranges, respectively. This relative velocity is also termed as the *range rate*.

The range rate varies as a function of three orbital factors, viz., satellite drift, inclination, and eccentricity,[52-53] using the subscripts d, i, and e, respectively, in the range-rate variation, $\frac{dR}{dt}$, we can write

Drift-Dependent Range-Rate Variation[54]

$$\left(\frac{dR}{dt}\right)_d = \frac{R_o A_o D}{2R_m} \cos\theta \sin(\Delta\phi) \qquad (11.11a)$$

where

R_o = earth's mean radius

A_o = major axis of the satellite

D = drift rate of the satellite with respect to the earth

R_m = mean range from the earth-station

θ = latitude of the earth-station

$\Delta\phi$ = difference in longitudes of the earth-station and the satellite

Inclination-Dependent Range-Rate Variation

$$\left(\frac{dR}{dt}\right)_i = \frac{R_0 A_0 wi}{2R_m} \sin\theta \cos(wt_a) \quad (11.11b)$$

where

$w =$ geocentric mean angular velocity of the satellite with respect to the earth-station

$= 2\pi/T_s$

$i =$ inclination of the satellite

$t_a =$ time from ascending node of the satellite

$T_s =$ orbital period of the satellite

and other parameters are as already defined above (see Equation 11.11a).

Eccentricity-Dependent Range-Rate Variation

$$\left(\frac{dR}{dt}\right)_e = \frac{A_0^2 w \sin(wt_p) e}{4R_m} \quad (11.11c)$$

where

$t_p =$ time from perigee of the satellite

$e =$ eccentricity of the satellite

and other parameters are as already defined above. Then the total range rate at any instant of time is given by the sum

$$\left(\frac{dR}{dt}\right)_T = \left(\frac{dR}{dt}\right)_d + \left(\frac{dR}{dt}\right)_i + \left(\frac{dR}{dt}\right)_e \quad (11.11d)$$

where the subscript T on the left designates the *total range-rate*.

The maximum value of the frequency shift, Δf_{max}, due to the Doppler effect can be approximately written as (CCIR Report 214):

$$\Delta f_{max} \approx \pm 3.0 \times 10^{-6} f N_d \quad (11.11e)$$

where f is the operating frequency, and N_d is the number of revolutions the satellite makes with respect to the earth-station (or with respect to a fixed point on the earth) in one day (24 hours).

11.5. Variation of the Satellite Apparent Position

Usually, a non-geostationary satellite appears to undergo an East-West (E-W) oscillation with respect to a given earth-station or fixed location on the earth. Two factors contribute to such an E-W oscillation: the satellite eccentricity (e) and the satellite inclination (i), and the amplitudes (maximum values) of the corresponding oscillations with respect to its nominal position are given as follows:

Eccentricity-Dependent Amplitude of the E-W Oscillation (A_e)

$$A_e I_{E-W} = 2e \text{ radians} \quad (11.12a)$$

Where

I_{E-W} is the maximum amplitude of the East – West oscillation

As an example, an eccentricity of 0.002 corresponds to amplitude of E-W oscillation equal to 0.004 radians, or $0.229°$ with respect to the nominal position of the satellite.

Inclination-Dependent Amplitude of the North-South (N-S) Oscillation (A_i)

For a circular orbit, the apparent N-S oscillatory motion occurs with the amplitude for a given satellite inclination as

$$A_i |_{E-W} = \sin^{-1}\left[\tan^2\left(\frac{i}{2}\right)\right] \qquad (11.12b)$$

$$\simeq \frac{1}{229} i^2 \qquad (11.12c)$$

where i is the value of the inclination in degrees.

The E-W oscillation has a period of half a day (12 hours). The period can be divided into four quarters, each of 3-hour duration. The apparent satellite drift, as part of the E-W oscillation, is westward during the first quarter and then eastward for the next two (second and third) quarters. Then the satellite again appears to move westward during the last (fourth) quarter, completing the full cycle of this E-W oscillation with respect to its nominal position.

11.6. Change in the Period of Revolution

With respect to a given earth station, a low-earth orbit (LEO) or medium-earth orbit (MEO) satellite, with circular geocentric orbit of certain radius $r_o < R_g$, R_g being the radius of a geostationary-earth orbit (GEO), moves faster as r_o decreases (lower orbit). A high-earth orbit (HEO) satellite, on the other hand, moves slower as its radius $r_o > R_g$ of the circular orbit increases. In other words, if the angular velocity of the satellite with respect to earth's center is w_o, and the angular velocity of a GEO satellite (i.e., the angular velocity of the earth) is $w_G (= w_E)$, where the suffices G and E refer to geostationary (GEO) satellite and the earth, respectively; and if R_G represents the altitude of a GEO satellite, so that for a satellite of altitude r_a, denoting the earth's radius as R_o, we can write the following inequalities

$$\text{LEO, MEO: } r_a < R_G \tag{11.13a}$$

$$\text{HEO: } r_a > R_G \tag{11.13b}$$

and the following equalities for the three types of satellites

$$\text{LEO, MEO, } r_o = r_a + R_o \tag{11.13c}$$

$$\text{HEO } r_a = r_o - R_o \tag{11.13d}$$

Note that an LEO satellite would drift in eastward direction with respect to a given earth-station; while an HEO satellite drifts westward.

Thus, the period and the orbital radius of the satellite would be different compared to that of the GEO satellite, and these variations can be related to the angular velocities and the orbital radius of the GEO satellite. The relative changes in these parameters, consistent with Kepler's Third Law, are given as follows; also given below is the change in the satellite velocity to correct the associated drift (to bring the satellite into the GEO orbit.)

$$\frac{\Delta T}{T_G} = \frac{\Delta w}{w_G} \tag{11.14a}$$

and

$$\frac{\Delta r}{R_g} = -\left(\frac{2}{3}\right)\frac{\Delta w}{w_G} \tag{11.14b}$$

and

$$\Delta v_c = \left(\frac{1}{3}\right)\frac{\Delta w}{w_G} \qquad (11.14c)$$

$$= \left(\frac{1}{3}\right) r_o \Delta w \qquad (11.14d)$$

where

Δw = change in angular velocity of the satellite

w_G = angular velocity of the GEO satellite (= angular velocity of the earth)

ΔT = change in the orbital period of the satellite

T_G = orbital period of a GEO satellite ($\simeq$ 24 hours)

ΔV_c = change in (linear) velocity of the satellite to correct for the drift

v_G = (linear) velocity of the GEO satellite

Also, recall that

r_o = orbital radius of the satellite

and

R_g = orbital radius of a GEO satellite

11.7. Mobile Systems Beam/Cell Coverage[54]

11.7.1. Introduction

Generally, the requirement of mobile communications system in a service area will determine the coverage pattern. Usually, the whole area is subdivided into cells, each cell being equipped with an antenna erected at a suitably chosen point ('cell-site') atop a pole-structure of a suitable height.

For designing the beam coverage pattern, the type of service area (metropolitan, urban, suburban, building structure, etc.) and also the natural terrain type (hilly, flat, foliage, water-body such as large lake, etc.) need to be taken into account. Other useful parameters involved are, of course, the height of the cell-site antenna, the distance between the cell-site center to the mobile unit, the ground incident angle (the angle between the direction of the incident electromagnetic (EM) wave and the ground plane at the mobile unit), the elevation angle (the angle between the EM wave and the horizontal plane at the mobile unit), the carrier frequency and the various sources of interference. The transmit power by the base station of the coverage and antenna gain values for the base unit and the mobile unit are additional EM parameters for a point-to-point model of the system.

Conventionally, for obtaining the EM path-loss in a selected area (city, suburban area, etc.) including man-made structures (e.g., clusters of buildings), the terrain pattern is ignored (i.e., a flat-terrain assumption made), and an "area-to-area" prediction curve for the path-loss as a function of distance between the transmitter and the mobile unit is obtained. This of course represents only the average variation pattern of the path-loss with distance. It is possible to represent this area-to-area[55-58] (A2A) average path-loss (in dB) against distance by a straight line.

11.7.2. Ground Reflection of the Incident Wave

At the point of incidence of the transmitted wave, the line-of-sight (LOS) path-loss and the ground-reflected wave power are given as follows:

$$P_L = P_t \left(\frac{\lambda}{4\pi D}\right)^2 \qquad (11.15a)$$

and

$$P_r = P_L \left|1 + C_r e^{i\Delta\phi}\right|^2 \qquad (11.15b)$$

where

P_t = transmitted power

λ = wavelength

D = distance between the transmitter and the mobile unit

P_L = LOS path-loss

P_r = ground reflected wave power

C_r = reflection coefficient

$\Delta\phi$ = phase difference between the incident and reflected wave

$\simeq \rho\Delta D$

ρ = phase different coefficient

ΔD = difference in the path-lengths of the incident and reflected waves

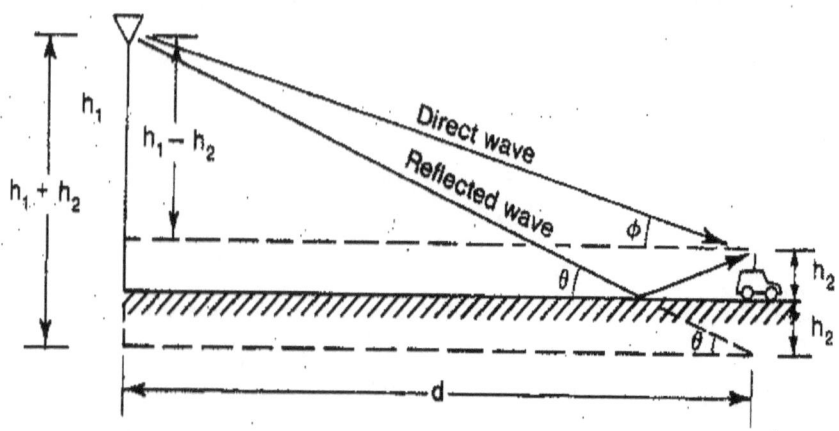

Figure 11.5. The Geometry of Ground Reflection of the Incident Wave [Adapted from William Y.S. Lee, *ibid*, p.357, by Permission from McGraw Hill].

Clearly, (see Figure 11.5),

$$\Delta D = D_1 - D_2 = [(h_1 + h_2)^2 + D^2]^{1/2} - [(h_1 - h_2)^2 + D^2]^{1/2}$$

where

D_1 = path-length of the incident wave

D_2 = path-length of the reflected wave

h_1 = height of the (cell-site) transmitter (antenna)

h_2 = height of the mobile unit antenna

Making the approximations, since $(h_1 \pm h_2 \ll D)$,

$$D_1 \simeq D\left[\left(\frac{h_1+h_2}{D}\right)^2 + 1\right]^{1/2} \simeq D\left[1 + \frac{1}{2}\left(\frac{h_1+h_2}{D}\right)^2\right]$$

and

$$D_2 \simeq D\left[\left(\frac{h_1-h_2}{D}\right)^2 + 1\right]^{1/2} \simeq D\left[1 + \frac{1}{2}\left(\frac{h_1-h_2}{D}\right)^2\right]$$

Therefore, we have

$$\Delta D \cong \frac{1}{2D}[2h_1h_2 - (-2h_1h_2)]$$

$$\simeq \frac{2h_1h_2}{D} \qquad (11.15c)$$

Also, assuming

$$\rho \simeq \frac{2\pi}{\lambda}$$

$$C_r \cong 1$$

and noting that

$$\Delta\phi \simeq \left(\frac{2\pi}{\lambda}\right) \times \frac{2h_1h_2}{D} = \frac{4\pi}{\lambda}h_1h_2/D$$

$$e^{i\Delta\phi} = \cos(\Delta\phi) + i\,\sin(\Delta\phi) \simeq 1 + i(\Delta\phi)$$

where we have used the approximations

$$\cos(\Delta\phi) \simeq 1,$$

$$\sin(\Delta\phi) \simeq \Delta\phi \qquad (\Delta\phi < 0.6 \text{ radian})$$

for small values of the angle ($\Delta\phi$), so that in Equation (11.15b), we can substitute

$$\left|1 + C_r e^{i\Delta\phi}\right|^2 \simeq \left|1 + \{(1 + i\sin(\Delta\phi)\}\right|^2$$

$$\simeq 4\left|1 + \frac{i\sin(\Delta\phi)}{2}\right|^2$$

$$\simeq 4\left(1 + \frac{\sin^2\Delta\phi}{4}\right)$$

$$\simeq 4\{1 + \frac{1}{4}(\Delta\phi)^2\}$$

$$\simeq 4\left[1 + \frac{1}{4}\left(\frac{4\pi\ h_1 h_2}{\lambda\ D}\right)^2\right]$$

As the second term on the right is much larger than 1 (since $\lambda \ll 1$), we have

$$\left|1 + C_r e^{i\Delta\phi}\right|^2 \simeq \left(\frac{4\pi\ h_1 h_2}{\lambda\ D}\right)^2,$$

Substituting in Equations (11.15a) and (11.15b), we get

$$P_L = P_t \left(\frac{\lambda}{4\pi D}\right)^2 \tag{11.16a}$$

and

$$P_r = P_L \left(\frac{4\pi}{\lambda}\frac{h_1 h_2}{D}\right)^2$$

$$\simeq P_t \left(\frac{\lambda}{4\pi D}\right)^2 \left(\frac{4\pi}{\lambda}\frac{h_1 h_2}{D}\right)^2$$

i.e.,

$$P_r \simeq P_t \left(\frac{h_1 h_2}{D^2}\right)^2 \tag{11.16b}$$

Consequently, in dB,

$$10\log_{10}(P_r) \simeq 10\log_{10}(P_t) + 20\log_{10}\left(\frac{h_1 h_2}{D^2}\right)$$

i.e.,

$$P_r(dB) \simeq P_t(dB) + 20\log\left(\frac{h_1 h_2}{D^2}\right)$$

Hence, the difference in the power levels ($\Delta P(dB)$) could be simply written as

$$\Delta P \cong P_t(dB) - P_r(dB)$$

$$\cong 20\log\left(\frac{D^2}{h_1 h_2}\right) \qquad (11.16c)$$

Noting that $h_1 h_2 \ll D$ and keeping only significant terms (neglecting second and higher order contributions), the difference in the power levels and antenna gains (at the transmitting base station and receiving mobile terminals) can be approximately evaluated.[59]

The signal strength (for the theoretical or measured curve) typically has a standard deviation of 8 dB.

If man-made structure and terrain contour are taken into account, appropriate modifications to the above results must be made. Usually, it is sufficient to treat EM propagation over water as equivalent to free space propagation. Man-made structure reduces the receive (or reflected) power, while terrain contour enhances it by the peaks of contour adding to it through scattered radiation. Thus, the modified or resultant reflected power could be expressed in the form

$$P_r = \{(P_{ro} - \beta \log \frac{r}{r_0}) + (20\log \frac{h_2}{h_1} + \propto)\} \quad ,$$

for non-obstructive path

$$\text{or, } P_r = \{(P_{ro} - \beta \log \frac{r}{r_0}) + (20\log \frac{h_2}{h_1} + \mathcal{L} + \propto)\} \quad ,$$

for obstructive path

$$(11.7)$$

where

P_{ro} = power received for free space (or over water) propagation

h_1 = height of the contour peak

r_o = path-length under freespace (or over water) propagation

r = path-length in the presence of man-made structure

and

$\alpha, \beta, \mathcal{L}$ = suitable adjustment parameters for diffraction losses

The above expressions are referred to as Lee Model[60]. Further simplification of this model can be realized by introducing the parameter v:

$$v = -h_k \left[\frac{2}{\lambda} \left(\frac{1}{r_1} + \frac{1}{r_2} \right) \right]^{1/2} \qquad (11.18)$$

where

h_k = height of the terrain contour peak relative to the height of the mobile unit

λ = wavelength

r_1 = distance of the contour peak to the cell-site antenna

r_2 = distance of the contour peak to the mobile unit

Depending on the terrain contour, h_k may be positive (when the peak height exceeds the height of the mobile unit) or negative (when the peak height is lower than the mobile unit height). For different ranges of the values of v, the diffraction loss ($\mathcal{L}$) can be estimated by a numerical fitting method. The result can be summarized as follows:

Range of v-values	Diffraction Loss $\mathcal{L}$ (dB)
$v \geq 1$	0
$0 \leq v < 1$	$20 \log(0.5 + 0.62v)$
$-1 \leq v < 0$	$20 \log(0.5 e^{0.95v})$
$-2.4 \leq v < -1$	$20 \log(0.4 - \sqrt{0.1184 - (0.1v + 0.38)^2})$
$-2.4 > v$	$20 \log\left(-\dfrac{0.225}{v}\right)$

In particular, we observe that for a terrain contour peak height equal to the mobile unit height, $v = 0$ and the diffraction loss reduces to b dB. Also, in case of more than one contour peak, the total diffraction loss can be obtained by adding the losses contributed by each peak, i.e.,

$$\mathcal{L} = \sum_{i} \mathcal{L}_i$$

where $\mathcal{L}_i$ is the diffraction loss due to the ith peak. Usually, it is sufficient to limit the summation on the right of the equation to two or three highest peaks. Within about 2 miles from the cell-site, the mobile unit can gain in the reflected power by up to about 10 dB compared to a flat terrain (i.e., the road on which the mobile unit is traveling is horizontal, i.e., perpendicular to the (vertical) cell-site antenna.

Cell Breathing

The mobile communication system commonly uses the CDMA access, where the transmissions by (or transmitted signals to) all other users act as interference noise. Thus, for reliable service normally available to a fixed number (say, N) CDMA mobile users, if the instantaneous number of users (say, N_i) increases ($N_i > N$), then the cell-size is suitably decreased to bring the number of users back to approximately equal to N. On the other hand, if the number of instantaneous users decreases ($N_i < N$), then the cell-size is appropriately increased so as to effectively make $N_i \simeq N$. This process of increasing and decreasing the cell-size due to the expected variations of the instantaneous number of mobile users is called *cell breathing*.

The cell breathing process may also come into play if one or more users (mobile units) are downloading a large data-file. In such a case, some of the voice users could be transferred ("soft handoff") to another cell, thereby necessitating an expansion of the cell-size of the original cell. As part of the cell-breathing process, any dead spot (no users) can also be eliminated.

11.7.3. System Efficiency

In digital cellular and mobile communications, the main technique is to use a multiplicity of CDMA carriers which use the same frequency bandwidth under *spread spectrum* or *frequency hopping* techniques, or by subdividing the total available bandwidth, which corresponds to subdividing the broadband carriers into many subcarriers – a technique known as *orthogonal frequency division multiple access* (OFDMA). In either case, the measure of the system efficiency is the number of voice channels that can be provided per unit bandwidth of the available spectrum. An estimate of the spectral efficiency of digital mobile/cellular systems is provided below, followed by a brief comparison thereof.

CDMA System Efficiency

The CDMA voice channels typically are based on voice encoder (vocoder) bit rates of 8 kbps or 13 kbps. In a 1.25 MHz of spread spectrum, a single CDMA carrier (cdmaOne) with 20-25 code channels can be accommodated with reasonable quality. The carrier with the combined CDMA signal is typically 2.4 Mbps capacity.

For the CDMA spreading function, a Walsh code is commonly used, typical code length being equivalent to 128 chips. Theoretically, this should yield 128 codes, thereby allowing 128 voice channels. However, due to the Walsh code not being a perfect code, the number of actual codes in practice is reduced substantially due to multiuser interference, especially in a multipath environment. Also, in this (CDMA) case, only soft handoff can be implemented. For those reasons, the 12- chip CDMA system can commonly provide only 20-25 code channels (i.e., allows only up to about 25 CDMA voice channels). Thus, a conservative estimate of the spectrum efficiency of CDMA can be written as[61]:

$$\frac{24\ channels}{2.4\ MHz} = 10\ (CDMA)\ channels/MHz$$

(11.10a)

For comparison, in an AMPS (frequency reuse) system with a seven-fold frequency reuse and with 3 sectors/cell, a total of 21 sectors results; and a 2 channel/cell channel efficiency (the term used for a frequency reuse system for system efficiency), would yield 2 x 21 = 42 channels, each effectively using a 30 kHz bandwidth, thereby implying a total bandwidth used equal to

$$42 \times 30\ kHz = 1260\ kHz = 1.26\ MHz \simeq 1.25\ MHz$$

(11.19b)

In other words, the CDMA system efficiency is about 10 times higher than that of the AMPS system.

If the multipath delay time is greater than the bit period of a narrow band digital channel, then the broad channel obtained by summing up the set of narrow band channels can provide the maximum data rate corresponding to the broadband channel.

However, reduction in the cdmaOne spectrum efficiency arises due to multiuser interference and general multipath condition (e.g., $\frac{30}{128} = 0.234$, i.e., about a 23% reduction); and also due to the soft handoff implementation (about 70%). The net reduction in capacity can therefore become 0.234 x 0.7 $\simeq$ 0.164; i.e, about 16%.

System Efficiency in Frequency Reuse Cellular System (FRCS): Radio Capacity[62]

The traffic capacity of a cellular system with frequency reuse is commonly expressed in terms of the radio capacity (ρ) defined as follows:

$$\rho = \frac{N}{k \times s} \qquad (11.20a)$$

where

$N =$ net number of channels

$k =$ frequency reuse factor (number of cells using the same frequency = the number of times the same frequency is used)

$s =$ number of 'sectors' into which a cell is divided in a sectored cell configuration (note: $s = 1$ in an omni-cell configuration)

The value of k depends on the geometry of the cellular system. The configuration of two different types of sector cells is illustrated in Figure 11.6.

If the separation between two nearest cells featuring frequency reuse is D, and the dimension of each cell is represented by an equivalent circular area of radius r, then the value of k and the carrier-to-interference noise ratio (C/I) could be related to the ratio, a.

$$a = \frac{D}{r} \tag{11.20b}$$

as follows

$$k = \frac{1}{3} a^2 \tag{11.20c}$$

$$\frac{C}{I} = \frac{1}{6} a^4 \tag{11.20d}$$

i.e.,

$$k = \frac{1}{3}\sqrt{6\frac{C}{I}} = \sqrt{\frac{2}{3}\frac{C}{I}} \tag{11.21a}$$

Combining Equations (11.20a) and (11.21a), we have

$$\rho = \frac{N}{s}\left(\frac{2}{3}\frac{C}{I}\right)^{-\frac{1}{2}} \tag{11.21b}$$

The above definition of the radio capacity (ρ) in a frequency reuse cellular system is useful in deriving various other traffic parameters of the system; viz., erlang/km^2, erlang/cell, calls/km^2, normalized (say/Hz) radio capacity. Clearly, the last parameter, normalized radio capacity (ρ_1) is given by

$$\rho_1 = \rho/B_t = \frac{N}{sB_t}\left(\frac{2}{3}\frac{C}{I}\right)^{-\frac{1}{2}} \tag{11.21c}$$

where B_t is the total (spectral) frequency band involved. The values of ρ_1 (in number of channels/cell/unit frequency band) provides a general estimate of the relative efficacies of two or more frequency reuse cellular systems.

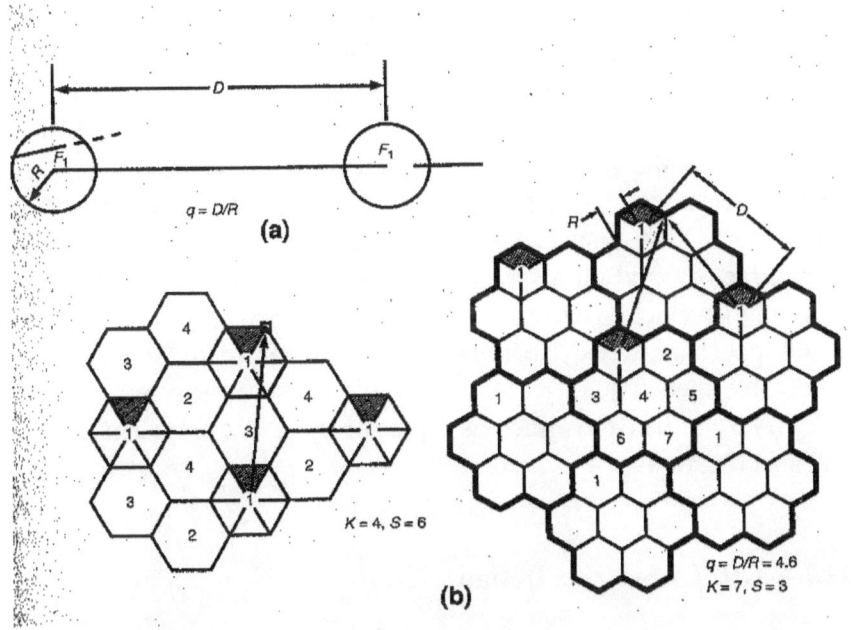

Figure 11.6. Two different types of cell sectors
[Adapted from Lee, *ibid*, p. 649, by Permission from McGraw Hill].

Processing Gain

In a CDMA system, the ratio of the spread bandwidth, B, and the bit rate, R_b, represents the processing gain (G_p). Each coded channel in CDMA is analogous to a frequency channel in an FDMA or TDMA, system with an important distinction: the channel noise level in the case of CDMA may be comparatively large – even larger than the signal power level. The CDMA channel noise of course is primarily due to the signal of other users using simultaneously the same channel (frequency band), and the signal of a particular user is still retrievable by virtue of its unique code.

$$G_p = B/R_b = BT_b \qquad (11.22)$$

where T_b is the bit period.

Apart from the processing gain, G_p, a most critical factor in a CDMA system is the ratio E_b/I_o, where E_b is the energy per bit, and I_o is the power density (per Hz). This energy-per-bit-to-power-density-ratio (E_b/I_o) determines the frame error rate (FER) at a given speed of the mobile user (vehicle). These factors (E_b/I_o, FER), in turn, affect the speech quality, which is of course also dependent on the voice-coding algorithm of a particular vocoder type employed. Thus, the CDMA system design must take into account these inter-related factors, and for a specified value or range of E_b/I_o, the full range of the vehicle speed and of the environmental conditions, as well as the speech activity cycle, must be properly considered to ensure an acceptable speech quality to the users. Conversely, the required level of E_b/I_o can be determined as a part of the CDMA system design and specifications.

11.7.4. CDMA Systems Design

The CDMA system Carrier-to-Interference-Noise Ratio, C/I, can be expressed in terms of other system parameters as follows:

$$\frac{C}{I} = \left(\frac{E_b}{I_o}\right)\left(\frac{R_b}{B}\right) \propto \qquad (11.23)$$

where $\propto$ is the speech activity cycle (in percent), and other parameters are as defined above.

Under different conditions, the frame error rate (FER) varies. The E_b/I_o value is suitably adjusted to meet a specified FER value for given environmental factors. The worst case scenario, together with any E_b/I_o specification then determines the required C/I value.

In a mobile CDMA system, the modulation scheme of the forward link channel is different from that of the reverse link channel.

Accordingly, the forward link C/I value, $(C/I)_F$, is different from the reverse link C/I value, $(C/I)_R$. The ratios, $(C/I)_F$ and $(C/I)_R$, are now derived separately.

Forward Link: $(C/I)_F$

The amount of interference noise in the forward link varies according to the distance or class of the interfering cells, as described below.

(a) Intracell Interference: The intracell value of contribution to $(C/I)_F$ can be simply written as $(C/I)_1$ given as

$$\left(\frac{C}{I}\right)_1 = \frac{P_o}{(N-1)P_o} = \frac{1}{(N-1)} \qquad (11.24a)$$

where, in the cell C_1 (primary cell),

P_o = power level of a single CDMA transmitter (= desired signal power) in the cell C_1

N = number of all the CDMA users in the cell C_1 (i.e., the number of users within C_1 acting as interference noise source = $N-1$).

$N = N_1$, (say)

(b) Adjacent Cell Interference

Corresponding to C_1, there are <u>two cells</u>, viz. C_2 and C_3, which can potentially generate interference noise for a forward link. Thus, their combined contribution to $(C/I)_F$ can be written as:

$$\left(\frac{C}{I}\right)_a = \frac{P_o}{P_2 N_2 + P_3 N_3} \qquad (11.24b)$$

where

P_i = the power level of each user in cell C_i;

N_i = number of CDMA user in cell C_i; $i = 2, 3$.

(c) Interim Cell Interference

There are <u>three cells</u> (C_4, C_5, and C_6) which are at an interim separations.

The combined contribution of these three interim cells can be written as:

$$\left(\frac{C}{I}\right)_i = \frac{P_o}{\beta(2)^{-4}} \qquad (11.24c)$$

where β is a constant (of the system)

(d) Distant Cell Interference

There are <u>six cells</u> ($C_7, C_8, \ldots C_{12}$) which contribute to the $(C/I)_F$. These cells are each at a distance equivalent to 2-cell separations.

$$\left(\frac{C}{I}\right)_d = \frac{P_o}{r(2.633)^{-4}} \qquad (11.24d)$$

where r is also a constant.

Total Interference

Combining the various contributions indicated above, the net forward link $\left(\frac{C}{I}\right)_F$ can be written as:

$$\left(\frac{C}{I}\right)_F^{-1} = \left(\frac{C}{I}\right)_1^{-1} + \left(\frac{C}{I}\right)_a^{-1} + \left(\frac{C}{I}\right)_i^{-1} + \left(\frac{C}{I}\right)_d^{-1}$$

(11.25a)

i.e.,

$$\left(\frac{C}{I}\right)_F^{-1} = (N_1 - 1) + \frac{P_2 N_2 + P_3 N_3}{P_0} + \frac{\beta}{P_0 2^4} + \frac{r}{P_0 (2.633)^4}$$

(11.25b)

Solving for N_1, the value of permissible number of mobile cellular users can be written as:

$$N_1 = 1 + \left(\frac{C}{I}\right)_F^{-1} - \frac{1}{P_0}\left[P_2 N_2 + P_3 N_3 + \frac{\beta}{2^4} + \frac{r}{(2.633)^4}\right]$$

(11.25c)

The following special cases may now be considered:

<u>Special Case 1</u> N_o (i.e., negligible) adjacent cell interference.

Setting for this case $P_2 = P_3 = \beta = r = 0$, we have

$$N_1 = 1 + \left(\frac{C}{I}\right)_F^{-1} \qquad (11.26)$$

For example, if

$$\left(\frac{C}{I}\right)_F = -20 dB,$$

i.e.,

$$\left(\frac{C}{I}\right)_F (dB) = 10 \log_{10} \left(\frac{C}{I}\right)_F = -20 dB$$

i.e.,

$$\log\left(\frac{C}{I}\right)_F = -\frac{20}{10} = -2.0 = -\log\left(\frac{C}{I}\right)_F^{-1}$$

or

$$\left(\frac{C}{I}\right)_F^{-1} = 10^2 = 100$$

Hence, from Equation (11.26), we get

$$N_1 = 1 + 100 = 101$$

Special Case 2

Interference other than the adjacent cells (i.e., the interim and distant cells) is negligible.

In this case, setting $\beta = r = 0$, we have from Equation (11.25c),

$$N_1 = 1 + \left(\frac{C}{I}\right)_F^{-1} - \frac{1}{P_0}[P_2 N_2 + P_3 N_3] \quad (11.27)$$

Assuming the power levels P_2 and P_3, and also the number of users in adjacent cells, N_2 and N_3, respectively, to be known, and also the value of $\left(\frac{C}{I}\right)_F$ to be known, we can determine either N_1 or P_0 if the other's value is known. Alternatively, the appropriate ratio of power levels to retain a certain value for N_1 can be obtained. For instance, if $N_1 = 101$, $\left(\frac{C}{I}\right)_F^{-1} = 100$ (as in the preceding special case, and also assuming $N_2 = 50$ and $N_3 = 40$, the above equation yields:

$$N_1 = 1 + 100 - \frac{1}{P_0}[50\,P_2 + 40\,P_3]$$

Since $N_1 = 101$, we obtain

$$5P_2 + 4P_3 = 0$$

or

$$\frac{P_2}{P_3} = -\frac{4}{5}$$

This gives the appropriate value of the ratio P_2/P_3 so as to retain the capacity of the cell C_1 unchanged.

Alternatively, if N_1 is treated as a variable, then, for $\left(\frac{C}{I}\right)_F^{-1}$ value to remain unchanged (100), we have:

$$N_1 = 101 - \frac{10}{P_0}[5P_2 + 4P_3]$$

i.e.,

$$5\left(\frac{P_2}{P_0}\right) + 4\left(\frac{P_3}{P_0}\right) = \frac{101 - N_1}{10}$$

which yields the appropriate ratios of the power levels P_0, P_2, P_3 to be satisfied for a given N_1-value.

It is easy to see that Equation (11.27) can also be expressed in the following forms; (writing $P_0 = P_1$).

$$N_1 P_1 + N_2 P_2 + N_3 P_3 = F N_1 \qquad (11.27b)$$
$$= F N_2$$
$$= F N_3$$
$$= F P_0$$

where

$$F = 1 + \left(\frac{C}{I}\right)_F^{-1}$$

Hence also, from Equation (11.27b), we have the equality

$$N_1 = N_2 = N_3 \qquad (11.27c)$$

The Reverse Link

Following a similar approach and using analogous notations for the reverse link, as for the forward link discussed in the preceding section (see Equation 11.24b), we can write the carrier-to-interference ratio, C/I, in the cell C_1 with a total of N_1 users (i.e., with $(N-1)$ CDMA users acting as sources of interference within C_1), and also limiting consideration to only the neighboring two cells C_2 and C_3 (i.e., assuming that the interim and distant cells cause negligible interference for the reverse link), we can write, for the home-cell (C_1),

$$\frac{C}{I} = \frac{P_0}{(N_1-1)P_0 + N_2 f_2 P_2 + N_3 f_3 P_3} \qquad (11.28a)$$

where

f_i = the fraction of the total number N_i of users in the adjacent cell C_i ($i = 2, 3$), and other parameters are as defined above.

Equation (11.28a) can be further simplified if we assume that all users in the CDMA network use the same transmit power (i.e., $P_2 = P_3 = P_0$), so that

$$\left(\frac{C}{I}\right)_R^{-1} = (N-1) + N_2 f_2 + N_3 f_3$$

For the worst case scenario, the following results are obtained:

$$1 + \left(\frac{C}{I}\right)_R^{-1} \geq N_1 + f_2 N_2 + f_3 N_3$$

or

$$\geq N f_1 + N_2 + f_3 N_3$$

or

$$\geq N f_1 + N_2 f_2 + N_3 \quad (11.28b)$$

where f_1 is the fraction of users in C_1 interfering with the desired signal. For design purposes, one may assume $f_1 = f_2 = f_3 = \frac{1}{6}$, (corresponding to one side of the hexagonal –shaped cells). For the CDMA network design, all the conditions embodied in equations or inequalities (11.28b) must be satisfied to ensure service with acceptable speech quality.

11.7.5. Antenna Coverage Pattern (ACP) for Cellular/Mobile (CEMO) Systems

Performance Considerations in Mobile Communications

For a successful operation of mobile telecommunications, a number of special considerations are involved, which are usually of little concern in connection with fixed satellite communication. These include the effect on the desired signal of propagation through buildings, co-channel interference due to emission from neighboring cells in a cellular network, the need for mobile-to-mobile communication, the need for directional antennas, and problems specific to a particular coverage design, etc. Now we turn our attention to some of these aspects of mobile communications.

Antennas for Cellular Communication

A number of antenna configurations may be employed for cellular communications in order to provide optimum coverage and performance for specific local conditions.

Co-Channel Interference

In order to determine the effect of co-channel interference, we first develop a general model for such an effect, and subsequently apply this model to the case of cellular communication towards the evaluation of pertinent carrier-to-interference ratio.

Let the desired signal, of frequency w, E_s, be represented by the sinusoidal function of amplitude $A_s(t)$, and phase ϕ_s,

$$E_s = A_s(t)\sin(wt + \phi_s) \qquad (11.29a)$$

while the interfering co-channel (frequency w) wave with a similar pattern of variation with amplitude $A_i(t)$ and phase ϕ_i, is representable as

$$E_i = A_i(t)\sin(wt + \phi_i) \qquad (11.29b)$$

where the time-variations of ϕ_s and ϕ_i are not explicitly indicated above for simplicity of notation; i.e., in actuality, $\phi_s \rightarrow \phi_s(t)$; $\phi_i \rightarrow \phi_i(t)$.

The total received signal is then represented by the sum (writing $A_s(t)$ and $A_i(t)$ as A_s and A_i) respectively,

$$E_T = E_s + E_i = A_s \sin(wt + \phi_s) + A_i \sin(wt + \phi_i)$$

$$= (A_s \cos\phi_s + A_i \cos\phi_i)\sin wt + (A_s \sin\phi_s + A_i \sin\phi_i)\cos wt$$

$$= A_T \sin(wt + \phi_T)$$

where

$$A_s \cos\phi_s + A_i \cos\phi_i = A_T \cos\phi_T \qquad (11.30a)$$

and

$$A_s \sin\phi_s + A_i \sin\phi_i = A_T \sin\phi_T \qquad (11.30b)$$

i.e.,

$$A_T^2 = [(A_s \cos\phi_s + A_i \cos\phi_i)^2 + (A_s \sin\phi_s + A_i \sin\phi_i)^2]$$

$$= [A_s^2 + A_i^2 + 2A_s A_i \cos(\phi_s - \phi_i)]$$

$$= F_1 + F_2 \qquad (11.31a)$$

and

$$\tan \phi_T = \frac{A_s \sin \phi_s + A_i \sin \phi_i}{A_s \cos \phi_s + A_i \cos \phi_i} \qquad (11.31b)$$

where

$$F_1 = A_s^2 + A_i^2 \qquad (11.31c)$$

and

$$F_2 = 2A_s A_i \cos(\phi_s - \phi_i) \qquad (11.31d)$$

Recalling that A_s, A_i, ϕ_s, and ϕ_i (and hence F_1 and F_2) are all varying or fluctuating with time, it is now useful to assess the mean (or statistical average) values of F_1 and F_2.

Clearly,

$$F_1(t) = A_s^2(t) * + A_i^2(t) * \qquad (11.32a)$$

where the stars by the symbols denotes a statistical average.

The statistical time-average of F_2 is best evaluated by first squaring it, so that

$$F_2^2(t) = 4A_s^2(t) A_i^2(t) \cos^2(\phi_s - \phi_i) *$$
$$= 2A_s^2(t) A_i^2(t) \qquad (11.32b)$$

where we have used the result $\cos^2 x * = \frac{1}{2}$ for any x.

Using Equation (11.32a), we can write the signal-to-interference ratio, R, in terms of the average power ratio, as

$$R = \frac{A_s^2(t)}{A_i^2(t)} \qquad (11.32c)$$

Combining Equations (11.32a), (11.32b), and (11.32c), we have

$$R = \frac{2F_1 A_s^2}{F_2^2} - 1 \qquad (11.33)$$

It can be shown[63] that $F_1(t)$ fluctuates close to the fading frequency V/λ, where V is the velocity and λ the wavelength of the signal wave; while the $F_2^2(t)$ fluctuates at a frequency close to $\left(\frac{d}{dt}(\phi_s - \phi_i)\right)$.

In practice, use can be made of an appropriately designed envelope detector (to obtain the statistical average values involved, an analog-to-digital converter, and a computational device (microcomputer). A processing delay time Δt may be involved due to these processing elements. It is assumed, however, that Δt is small enough to satisfy the conditions:

$$A_s(t + \Delta t) \simeq A_s(t) \qquad (11.34a)$$

$$A_i(t + \Delta t) \simeq A_i(t) \qquad (11.34b)$$

$$\cos[\phi_s(t + \Delta t) - \phi_i(t + \Delta t)] \simeq \cos[\phi_s(t) - \phi_i(t)]$$

$$(11.35)$$

On the other hand, Δt must be large enough to permit the necessary processing. Due to these conflicting requirements, a real-time evaluation of the co-channel signal-to-interference (power) ratio becomes a difficult task.

11.7.6. Spectral Density for Mobile-to-Mobile Communication[63]

11.7.6.1. Introduction

In most cases of mobile-to-mobile communication links, the transmitting terminal and the receiving terminal – typically each of them being a hand-held device with a passenger of a moving vehicle – are both in a state of motion. Clearly, both the transmitted and received waves entail Doppler effect. Furthermore, there may be a myriad of buildings, vegetation, and other obstacles in the path of propagation of the signal. Under the foregoing factors and circumstances, it is useful to represent the propagation path in terms of a transfer function. In this section, we derive the form of such a transfer function, and then apply the result to characterize the mobile-to-mobile transmission for analyzing the performance characteristics of such a link.

11.7.6.2. The Transfer Function

Consider two mobile units (terminals), u_1 and u_2, in a cellular or mobile network in motion, with velocities V_1 and V_2, respectively, in arbitrary directions. Let u_1 be the transmitting terminal; so the receiving terminal u_2 receives the transmitted signal from u_1 over a multiplicity of paths due to the presence of buildings and other obstacles between u_1 and u_2.

Let the transmitted signal, A_t, be represented as a plane wave, with amplitude $u(t)$ and frequency w, at time t:

$$A_t = u(t)e^{iwt} \qquad (11.36)$$

The received signal is a combination or sum of waves propagating in various paths. The signal received via the jth path, P_j can be written as

$$R_j = r_j u(t - \theta_j) \exp\left[i(w_0 + w_{1j} + w_{2j})(t - \theta_j) + \phi_j\right]$$

$$= S_j u(t - \theta_j) e^{iw_0(t - \theta_j)} \qquad (11.37)$$

where w_0 = RF carrier frequency

r_j = the amplitude factor of the wave via P_j

θ_j = the propagation time delay for P_j

ϕ_j = phase (a uniformly distributed random variable) of the wave along P_j

$w_{1j} = \frac{2\pi}{\lambda} V_1 \cos(\beta_{1j})$ = Doppler shift of the transmitted wave along P_j

$w_{2j} = \frac{2\pi}{\lambda} V_2 \cos(\beta_{2j})$ = Doppler shift of the received wave along P_j

β_{1j}, β_{2j} = random angles involved in the transmitted and received waves along P_j (due to the presence of buildings and other obstacles)

$$S_j = r_j \exp\left[i(w_{1j} + w_{2j})t + \phi_j^1\right] \qquad (11.38a)$$

with

$$\phi_j^1 = \phi_j - (w_{1j} + w_{2j})\theta_j \qquad (11.38b)$$

It should be noted that r_j is a random variable distributed according to the Rayleigh distribution; while the phase ϕ_j, is a randomly distributed variable. The angles β_{1j} and β_{2j} are also randomly distributed. The Doppler shifts are to be taken algebraically, i.e., with positive (for approaching) or negative (for receding) signs for the mobile unit.

In Equation (11.33), we have used the symbol $u(t - \theta_j)$ to represent the time delay along the jth path, P_j. For simplicity, let us now assume

$$u(t) = e^{iwt}$$

i.e.,

$$u(t - \theta_j) = e^{iw(t-\theta_j)}$$

Then, from Equation (11.33), we have

$$R_j = S_j \, e^{i(w+w_0)(t-\theta_j)} \tag{11.39}$$

The total signal received can then be represented by summing up the contributions of all such contributions; i.e., over all likely or effective paths $P_{i,j} = 1, 2, \ldots, n$.

Therefore, the total received signal R, can be simply written as:

$$R = \sum_{j=1}^{n} R_j = \left[\sum_{j=1}^{n} S_j \, e^{-i(w+w_0)\theta_j} \right] e^{i(w+w_0)t}$$

$$= T_{jw} \, e^{i(w+w_0)t} \tag{11.40a}$$

where

$$T_{jw} = \sum_{j=1}^{n}[S_j \, e^{-i(w+w_0)\theta_j}] \qquad (11.40b)$$

is the requisite transfer function.

If $w = 0$ (only sinusoidal carrier frequency), then we have the effective magnitude of the received wave amplitude as (see Equation (11.40a),

$$R_0 = |T_{jw}| \qquad (11.41)$$

R_0 is also a Rayleigh-distributed random variable with probability density function given as

$$R(R_0) = 2\,R_0\,P_0\,\exp(-R^2{}_0 P_0)$$

and with average power $1/P_0$.

11.7.6.3. Spectral Distribution

The spectral distribution of a time-dependent wave is obtained by Fourier transformation of the temporal autocorrelation function representing the wave. Thus, for obtaining the spectral distribution in question, we need to first determine the time-based autocorrelation function of the wave.

Autocorrelation Function[63]

Let the received signal envelope at time t_1 and point denoted by x_1 be given by

$$R(x_1, t_1) = \sum_{j=1}^{n} S_j \exp i[(w_{1j} + w_{2j})t + \theta_j + kx_1 \cos \alpha_{1j}]$$

(11.42)

where $k = 2\pi/l$.

The wave amplitude at the point x_2 at time t_2 is also expressible by a similar equation.

Therefore, the requisite spatial time correlation function is given by

$$R_{12}(x_1, x_2, t_1, t_2) = \frac{1}{2} < [R(x_1, t_1) R^*(x_2, t_2)] >$$

(11.43)

where the star (*) represents the complex conjugate.

For small differences in spatial and time variables

$$x_1 - x_2 = \Delta_x \qquad (11.44a)$$

$$t_1 - t_2 = \Delta_t \qquad (11.44b)$$

we can write the *time* correlation function for random stationary R-functionsas

$$R_{12} = (\Delta_t) = \alpha^2 J_0(\beta V_1 \Delta_t) J_0(\beta V_2 \Delta_t) + \beta \Delta_x)$$

(11.45a)

Where

α^2 = half the average power = $1/(2P_0)$

$\beta = 2\pi\lambda$

$J_0(\xi)$ = zero-order Bessel function of argument ξ

Similarly, the *spatial* correlation function is given as:

$$R_{12}(\Delta_x) = \alpha^2 J_0(\beta \Delta_x) \qquad (11.45b)$$

By taking the Fourier transformation of the correlation function, $R(\Delta_t)$, we obtain the spectral density $S(w)$

$$S(w) = \int_{-\infty}^{\infty} R(\Delta_t) \, e^{-iw\Delta_t} \, d(\Delta_t)$$

$$= \frac{2\alpha^2}{\pi \sqrt{w_1 w_2}} K_1 \left[\frac{\left(1+\frac{w_2}{w_1}\right)}{2\sqrt{\frac{w_2}{w_1}}} \sqrt{1 - \left\{\frac{w}{w_1+w_2}\right\}^2} \right]$$

(11.46a)

where

$$w_i = V_i k_i, \quad i = 1, 2$$

$$k_i = \frac{2\pi}{\lambda_i}$$

and

$K_1(\xi)$ = complete elliptic integral of the first kind and argument ξ.

If $V_2 = 0$, i.e., $w_2 = 0$, then the spectral density function reduces to

$$S(w) = \frac{4\alpha^2}{(w_1^2 - w^2)^{1/2}} \qquad (11.46b)$$

Equation (11.46b) represents the spectral density function for the cell-site (base)-to-mobile channel.

11.7.6.4. The Link Equation for a Mobile Receiver

In a mobile system network, the cell site may be judiciously chosen to optimize the link performance; but the location of the receive terminal may vary over a range of environmental conditions. The variation could include, for example, the receiver being in the open space without any obstacles or hindrance, or outside a building that stands in the way, or inside the building in a room.

The transmit power and transmit antenna gain remain unaltered in all such cases, and the G/T-value of the receive system is also assumed to remain identically the same. Thus, the only factor that undergoes variation in different scenarios of the above type is the path-loss. The appropriate path-loss can then be used in the link equation to determine the resultant performance. Also, of course, a combination of the above mentioned circumstances may be presented, including the case when all the three factors are present simultaneously.

In this Section, we provide expressions for the path-loss corresponding to main variation in the type of path or intervening obstacles, as exemplified above.

Case 1: Propagation in Free Space (No Obstacles)

In this case, the path-loss is simply as given earlier (Equation 11.15), except for any additional loss due to environmental conditions that also contribute to the path-loss. Such conditions may vary from place to place, depending on the level of industrial emissions (smoke or other gases) or polluting elements that change the quality of air or atmosphere through which the radiated wave carries the signal. The atmospheric absorption of wave energy would therefore, be different in a city in comparison to a suburban area. There may be marked variation from one city to another, as well.

In review of the above considerations, the propagation loss in a free space or open path (without natural or structural obstacles) can be generally written as

$$L_1 = L_0 + L_e \qquad (11.47a)$$

where L_0 is the usual distance-dependent path-loss, and L_x is the environment-dependent path-loss, i.e.,

$$L_0 = \left(\frac{4\pi D_1}{\lambda}\right)^2 \qquad (11.47b)$$

where D_1 is the distance between the transmitter and receiver, and λ is the carrier wavelength.

In dB, the open-space path-loss can be written as

$$L_1(dB) = 10\log\left(\frac{4\pi D_1}{\lambda}\right)^2 + L_e^1$$

$$= 20\log\left(\frac{4\pi D_1}{\lambda}\right) + L_e^1 \qquad (11.47c)$$

where $L_e^1 = 10\log L_e$ is the environment-dependent path-loss in dB, and log denotes logarithm with respect to base 10.

The value of L_e^1 for suburban areas and for a set of representative cities (i.e., New York, Philadelphia, Newark, Tokyo), based on actual measurements are available. L_e^1 may also include multipath-dependent loss due to uneven terrain (natural obstacles).[64-65]

Case 2: Receiver Inside a Room

If the receiver is inside a room, the total path-loss is the sum of the above (free-space) path-loss and the loss occurring due to the wave penetrating the intervening wall and due to further propagation from the wall to receiver. The total path-loss in this case can therefore be written as

$$L_2(dB) = L_1(dB) + \alpha_1 \log\left(1 + \frac{D_2}{D_1}\right) \qquad (11.47a)$$

where $L_1(dB)$ is as given by Equation (11.47c), D_2 is the distance between the intervening wall of the room and the receiver, and α_1 denotes the attenuation slope with value depending on the structural material of the wall. In the case of a 'regular' wall (typically made of bricks or concrete), the value of α_1 is approximately 40 dB/dec (dec here is abbreviation of decimeter); while for a 'special' room (such as an elevator or utility room), the value is much lesser. Note that if the intervening wall is not there (i.e., $D_2 = 0$), Equation (11.47d) reduces to Equation (11.47c), as expected.

Case 3: Receiver Outside a Building

If the receiver is outside a building such that the wave travels through a distance D_1 in free space, then it traverses a building spanning a distance D_2, then, emerging from the building, it travels an additional distance D_3 without further obstacle (i.e., in free space, again), then the total path-loss can be written as

$$L_3 = L_2 + L_w + 20\log\left(1 + \frac{D_3}{D_1 + D_2}\right) \quad (11.47e)$$

where L_w is the loss associated with the nature of the walls of the intervening building, D_2 in this case is the distance traversed *within* the building, and D_3 is the distance traversed between the wall from which the wave finally emerges and the location of the receiver. If $D_3 = 0$, then the Third term on the right of Equation (11.47e) vanishes and $L_3 = L_2 + L_w$. The value of L_w is typically in the range of (18 ± 3) dB depending on the type of structural material of the wall involved.

References:

[49] C. A. Sicoco, Broadcasting Satellite Coverage – Geometrical Considerations, IEEE Trans. Broadcasting, Vol. BC-19, No. 4, December 1973, pp. 84-87.
[50] Dr. C. Beste, Design of Satellite Constellation for Optimal Continuous Coverage, IEEE Trans. Aerosp. Electr. Systems, Vol. AES-14, No. 3, May 1978, pp. 466-473.
[51] G. Maral, J-JD. Ridder, B.G. Evans, and M. Richharia, Low Earth Orbit Satellite Systems for Communications.

International J. of Satellite Communications, Vol. 9, 1991, pp. 209-225.

[52] V.J. Slabinski, Variation in Range, Range-Rate, Propagation Time Delay and Doppler Shift in a Nearly Geostationary Satellite, Progress. In Astronautics and Aeronautics, Vol. 33, No. 3, 1974.

[53] W.I. Morgan and G.D. Gordon, Communications Satellite Handbook, John Wiley, New York, 1989. CCIR Report 214.

[54] Richharia, *ibid*, p. 430.

[55] K. Bullington, Radio Propagation for Vehicular Communications, IEEE Trans. On Vehicular Technology, Vol. VT-26, No. 4, 1977, pp. 295-308.

[56] H. Susuki, A Statistical Model for Urban Radio Propagation, IEEE Trans. On Communications, Vol. Com 25, July 1977.

[57] D.L. Nielson, Microwave Propagation Measurements for Mobile Digital Radio Applications, IEEE Trans. On Vehicular Technology, Vol. VT-27, August 1978, pp. 117-132.

[58] W.C.Y. Lee, Mobile Communications Engineering, Chapter 4, McGraw Hill, New York, 1998.

[59] W.C.Y. Lee, Mobile Communications Engineering, Chapter 4, McGraw Hill, New York, p. 358, 1998.

[60] W.C.Y. Lee, Lee's Model, IEEE VTS 42 Conference Proceedings, Denver, May 10-13, pp. 343-348, 1992.

[61] S. Kozono and M. Sakamoto, Channel Interference Measurement in Mobile Radio System, Proc. Of the 35th IEEE Vehicular Technology Conference, Boulder, CO, May 21-23, 1985, pp. 60-66.

[62] Lee, *ibid*, p. 391.

[63] Lee, *ibid*, p. 393.

[64] W.C.Y. Lee, Mobile Communications Engineering, McGraw Hill, New York, p. 125, 1998.

[65] W.C.Y. Lee, Wireless and Cellular Communications, McGraw Hill, New York, p. 354-357, 2006.

12. EPILOG: SATELLITE TELECOMMUNICATIONS AND BASIC PHYSICS

We conclude this book by noticing an intriguing perspective of satellite telecommunications: its symbiotic relationship with physics, one of the most basic branches of natural sciences. More specifically, satellite telecommunications has had, almost from the very start, a deep interconnection with elementary particle physics, general relativity, and cosmology. A few illustrative examples are presented below.

In mid-1960s, the Bell Laboratory in New Jersey, USA, was engaged in pursuing a certain engineering project in satellite communications for which two members of its staff, A.A. Penzias and R. W. Wilson, were working on an antenna to get it ready for some measurements of its performance. To determine the efficacy of the antenna, they needed to eliminate all extraneous sources of noise. Even with repeated effort they were not able to get rid of a peculiar hissing noise. It did not matter which direction the antenna was facing, this noise was insistently present. They even cleaned the antenna of some "white dielectrics" (bird-droppings) which they thought could be the cause of the 'noise' in question; yet the noise would just not go away.

A few miles away, in Princeton, New Jersey, a group of theoretical researchers led by the renowned cosmologist, Robert Dicke, was theoretically investigating the probable consequences of the so-called *big bang theory* about the origin of the universe. In combination with the observed expansion of the universe as evidenced by the measurements of the velocities of galaxies as a function of their distances, it can be concluded that for a big bang-generated universe, there would exist a *black body radiation* throughout the universe which, originally of the value of about 1029 K at the time of big bang, would have decayed by now to yield a black body radiation -- after some 14 billion years of expansion and radiation cooling – of approximately 3 K.

A chance meeting between the two groups (Penzias and Wilson on experimental measurement side on one hand, and Dicke on theoretical calculations side, on the other) essentially led to a conclusion and confirmation that what Penzias and Wilson accidentally observed was just the residual cosmological radiation characteristic of 3 K black body radiation. A pair of papers published by the two groups side-by-side in the Astrophysical Journal and follow-up measurements [66-68] led to convincing establishment of the big bang theory. A rival theory – the *steady state theory* by Fred Hoyle of U.K. – proposing that there was no such big bang origin for the universe; the universe existed in a steady state all the time, with continuous creation of matter to fill the gap from the Hubble expansion – was finally rejected.

Another example can be cited in connection with the Global Positioning System (GPS) that is now routinely used by mobile users and others to determine their location on the earth (in terms of latitude and longitude) most accurately. To achieve this exquisite accuracy, the GPS system takes into account a *general relativistic* effect which states that clocks run slower in a gravitational field. Unless the pertinent corrections are made in satellite timings, accuracy provided by the GPS system would not be possible.

As an additional example of intimate interrelation between satellite telecommunications and scientific applications, particularly in the fields of *elementary particle physics*, one may consider the *Large Hadron Collider (LHC)* at the European Center for Nuclear Research (CERN) in Geneva, Switzerland. Satellite-based Internet system is an indispensable part of CERN for data exchange and analysis among scientists and engineers across the globe. Recently (July 4, 2012) CERN announced that they have discovered the *Higgs boson* (the so-called *God Particle*). This discovery establishes with greater confidence the so-called *Standard Model* of elementary particle physics.

Satellites including the *Hubble Telescope* have helped a great deal in characterizing special phenomena including the uniform structure of the universe on a large scale, *inflation*[69], formation of the galaxies, the *entanglement* (termed by Einstein *"Spooky action at a distance"* larger than covered under the velocity of light), relative abundance of small nuclei in the universe, providing further support to the big bang theory, and so on.

Satellite-based cosmological observations, starting from the *Cosmic Background Explorer (COBE)* of NASA, and continuing with the *Hubble Telescope* and the *PLANCK satellite* of Europe, mark special interrelation between satellite telecommunications and particle physics and cosmological research and observations. This interrelation is destined only to grow with coming years and decades.

References:

[66] R.H. Dicke, P.J.E. Peebles, P.G. Roll, and D.T. Wilkinson, Astrophysical Journal, Vol. 142, pp. 414-419, 1965.
[67] A.A. Penzias and R.W. Wilson, Astrophysical Journal, Vol. 142, pp. 419-421, 1965.
[68] A.A. Penzias and R.W. Wilson, "A Measurement of the Background Temperature at 1415 MHz," Astrophysical Journal, Vol. 72, p. 315, 1967.
[69] Alan H. Guth, The Inflationary Universe, Perseus Books, Cambridge, Mass, 1997.

LIST OF REFERENCES

CHAPTER 1

[1] W. L. Pritchard, H. G. Suyderhoud, and R. A. Nelson, Satellite Communication Systems Engineering, Prentice Hall, Englewood, N.J. 1993.

[2] B. N. Agrawal, Design of Geosynchronous Spacecraft, Prentice Hall, Englewood Cliffs, NJ, 1986.

CHAPTER 3

[3] www.satsig.net/linkbugt.htm (This public-domain website also has links to calculate antenna pointing, VSAT-related calculations, etc.)

CHAPTER 5

[4] William Y. S. Lee, Wireless and Cellular Telecommunications, McGraw Hill, p. 95, 2006.

[5] William Y. S. Lee, Wireless and Cellular Telecommunications, McGraw Hill, p. 101, 2006.

[6] M. Richharia, Satellite Communications Systems, McGraw Hills, 1999, p. 250.

[7] William Y. S. Lee, Wireless and Cellular Telecommunications, McGraw Hill, p. 131, 2006.

[8] W. L. Pritchard et al, Satellite Communication Systems Engineering, Prentice Hall, p. 389, 1993.

[9] M. Richharia, Satellite Communications Systems, McGraw Hills, 1999, p. 391, 1999.

[10] D.V. Sarvate and M.B. Pursley, Cross Correlation Properties of Pseudo- Random and Related Sequences, Proc-IEEE, Vol. 68, p. 593-619, 1980.

CHAPTER 6

[11] C.E. Shanon, "A Mathematical Theory of Communications," Bell System Technical Journal (BSTJ), vol. 27, p. 379-623.
[12] R.W. Hamming, Coding and Information Theory, Prentice Hall, Englewood Cliff, NJ, 1980.
[13] P.G. Farrel and A.P. Clark, Modulation and Coding, International Journal of Satellite Communications, Vol. 2, 1984, pp. 287-304.
[14] J. A. Heller and I.M. Jacobs, Viterbi Decoding for Satellite and Space Communications, IEEE Trans. Communication Technology, COM-19, October 1971, pp. 835-848.
[15] W.W. Wu, Applications of Error-Coding Techniques to Satellite Communications, Comsat Technical Review, Vol. 1, Fall 1971, pp. 183- 219.
[16a] Richard W. Hamming, Coding and Information Theory, Prentice Hall, Englewood Cliff, NJ, p. 206, 1980.

CHAPTER 7

[16b] M. Richharia, Satelite Communications Systems, McGraw Hill, p. 183. 1999.
[17] Richard W. Hamming, Coding and Information Theory, Prentice Hall, 1980.
[18] C. Berrou and A. Glavieux, Near-optimum Error Correcting Coding and Decoding: Turbo-Codes, IEEE Trans. Communication, Vol. 44, No. 10, pp. 1261-1271. 1996.
[19] C. Berrou, A. Glavieux, and P. Thitimajshima, Near Shanon Limit Error Correcting Coding and Decoding: Turbo Codes, Proceedings of ICC '93, Geneva, May 1993, pp. 1064-1070.
[20] R.M. Tanner, A Recursive Approach to Low Complexity, IEEE Trans. Info., Theory, Sept. pp.533-547, 1981.

[21] M. Luby et al., Improved Low-Density Parity Check Codes Using Irregular Graphs, IEEE Trans. Info., Theory, Feb. 2001.

[22] T. Richardson and R. Urbanke, Effective Encoding of Low-Density Parity Check Codes, IEEE Trans. Info., Theory, Vol. 47, pp.638-656, 2001.

[23] Willian C.Y. Lee, Wireless and Cellular Telecommunications, p.744, 2006.

[24] A.J. Viterbi, Convolution Codes and Their Performance in Communication Systems, IEEE Trans. Communication Technology, COM-19, October 1971, pp. 751- 772.

[25] J.A. Heller and I.M. Jacobs, Viterbi Decoding for Satellite and Space Communication, IEEE Trans. Communication Technology, COM-19, October 1971, pp. 835-848.

CHAPTER 9

[26] William C.Y. Lee, Wireless and Cellular Telecommunications, McGraw Hill, 2006.[Note-It must be thankfully acknowledged that Reference 26 (W.C.Y.Lee) served to the author (A.K.Sinha) of this book as a primary Reference in the preparation of many segments of this book. A number of Figures, data, and Tables have been used in the present book from Reference 26, by Permission from the Publisher: McGraw Hill, as mentioned appropriately in the text of this book].

[27] S. Hartford, M. Webster, and J. Zyren, "CCK-OFDM Normative Text Summary," IEEE 802.11-01/436 rl, July 2001.

[28] Intel White Paper, "IEEE 802.16 and WiMAX Broadband Wireless Acess for Everyone," http://grouper.ieee.org/groups/802/16.

[29] C.Eklund, R.B. Marks, K.I. Stanwood, and S. Wang, "IEEE Standard 802.16: A Technical Overview," IEEE Communication Magazine, June 2002.

[30] Daniel Minoll, Hot Spot Networks, Wi-Fi for Public Access Locations, McGraw Hill, 2002.

[31] Daniel Minoll, Hot Spot Networks, Wi-Fi for Public Access Locations, McGraw Hill, 2002.
[32] T. Ojanpera and R. Pasad, Wideband CDMA for Third Generation Mobile Communications, Artech House Publishers, Boston, 1998 [Chapter 5 CDMA Air Interface Design].
[33] V.K. Garg, IS-95 CDMA and cdma2000, Prentice Hall, PTR, 2000.
[34] V. Vangi, A. Damnjanovic, and B. Vojcic, The cdma2000 System for Mobile Communications, Prentice Hall PTR, 2004.
[35] S.C. Yang, 3G cdma2000, Artech House, Boston, 2004.
[36] B. Pelletier and H. Leib, "PCS Third Generation CDMA System, Study of the Physical Layer," Wireless Communication Group at McGill University, Canada, August 2004.
[37] H. Holma and Antti Toskala, WCDMA for UMTS, John Wiley & Sons, 2001.
[38] C. Smith and D. Collins, 3G Wireless Works, McGraw Hill, 2002.
[39] J.S. Lee and L.E. Miller, "CDMA System Engineering Handbook," Artech House, Boston, 1998.
[40] 3GPP2, "cdma2000 Standard for Spread Spectrum Systems, Revision C, May 2002.
[41] 3GPP2, "cdma2000 Standard for Spread Spectrum Systems, Revision D," February 2004.
[42] The participating companies in this Forum included Samsung (Host), Nokia, Motorola, Ericson, Siemens, Wireless Broadband (WiBro), NTT McCoMo, etc.
[43] W.C.Y. Lee, CS-OFDMA: A New Wireless CDD Physical Layer Scheme, IEEE Communication Magazine, Vol. 43, February 2005, p.74-49.

CHAPTER 10

[44] H. Taub and D.L. Schilling, Principles of Communication Systems, McGraw Hill, New York, 1971.
[45] Pritchard et al, *ibid*, p. 347

[46] W.C.Y. Lee, Mobile Communications Design Fundamentals, Howard W. Sams, Indianapolis, Minnesota, 1986.
[47] A.K. Sinha, "A Model for Burst Assignment and Scheduling," COMSAT Technical Review, Vol 6, No. 2, P. 219, Fall 1976.
[48] Pritchard et al, *ibid*, p. 354.

CHAPTER 11

[49] C. A. Sicoco, Broadcasting Satellite Coverage – Geometrical Considerations, IEEE Trans. Broadcasting, Vol. BC-19, No. 4, December 1973, pp. 84-87.
[50] Dr. C. Beste, Design of Satellite Constellation for Optimal Continuous Coverage, IEEE Trans. Aerosp. Electr. Systems, Vol. AES-14, No. 3, May 1978, pp. 466-473.
[51] G. Maral, J-JD. Ridder, B.G. Evans, and M. Richharia, Low Earth Orbit Satellite Systems for Communications. International J. of Satellite Communications, Vol. 9, 1991, pp. 209-225.
[52] V.J. Slabinski, Variation in Range, Range-Rate, Propagation Time Delay and Doppler Shift in a Nearly Geostationary Satellite, Progress. In Astronautics and Aeronautics, Vol. 33, No. 3, 1974.
[53] W.I. Morgan and G.D. Gordon, Communications Satellite Handbook, John Wiley, New York, 1989. CCIR Report 214.
[54] Richharia, *ibid*, p. 430.
[55] K. Bullington, Radio Propagation for Vehicular Communications, IEEE Trans. On Vehicular Technology, Vol. VT-26, No. 4, 1977, pp. 295-308.
[56] H. Susuki, A Statistical Model for Urban Radio Propagation, IEEE Trans. On Communications, Vol. Com 25, July 1977.
[57] D.L. Nielson, Microwave Propagation Measurements for Mobile Digital Radio Applications, IEEE Trans. On Vehicular Technology, Vol. VT-27, August 1978, pp. 117-132.
[58] W.C.Y. Lee, Mobile Communications Engineering, Chapter 4, McGraw Hill, New York, 1998.

[59] W.C.Y. Lee, Mobile Communications Engineering, Chapter 4, McGraw Hill, New York, p. 358, 1998.

[60] W.C.Y. Lee, Lee's Model, IEEE VTS 42 Conference Proceedings, Denver, May 10-13, pp. 343-348, 1992.

[61] S. Kozono and M. Sakamoto, Channel Interference Measurement in Mobile Radio System, Proc. Of the 35th IEEE Vehicular Technology Conference, Boulder, CO, May 21-23, 1985, pp. 60-66.

[62] Lee, *ibid*, p. 391.

[63] Lee, *ibid*, p. 393.

[64] W.C.Y. Lee, Mobile Communications Engineering, McGraw Hill, New York, p. 125, 1998.

[65] W.C.Y. Lee, Wireless and Cellular Communications, McGraw Hill, New York, p. 354-357, 2006.

CHAPTER 12

[66] R.H. Dicke, P.J.E. Peebles, P.G. Roll, and D.T. Wilkinson, Astrophysical Journal, Vol. 142, pp. 414-419, 1965.

[67] A.A. Penzias and R.W. Wilson, Astrophysical Journal, Vol. 142, pp. 419- 421, 1965.

[68] A.A. Penzias and R.W. Wilson, "A Measurement of the Background Temperature at 1415 MHz," Astrophysical Journal, Vol. 72, p. 315, 1967.

[69] Alan H. Guth, The Inflationary Universe, Perseus Books, Cambridge, Mass, 1997.

LIST OF THE VALUE OF SOME IMPORTANT CONSTANTS*

Velocity of Light (in Vacuum)	$c = 2.9979246 \times 10^{10}$	cm/sec
Planck's Constant (Unit of Action)	$h = 6.626 \times 10^{-27}$	erg-sec
Electron Charge (Unit of Charge)	$e = (-)1.602 \times 10^{-19}$	Coulomb
Boltzmann Constant	$k = 1.380 \times 10^{-16}$	erg/K
Gravitational Constant	$G = 6.672 \times 10^{-8}$	cm^3/(s^{2}gm)
"Standard" Acceleration of Gravity	$g = 980.665$	cm/sec^2
"Standard" of Time (Cesium Clock)	$v = 9192631770$	Hz
Mass of the Sun	$M_s = 1.989 \times 10^{33}$	gm
Mass of the Earth	$M_E = 5.974 \times 10^{27}$	gm
Mass of the Moon	$M_M = 7.348 \times 10^{25}$	gm
Eccentricity of the Earth	$e_c = 0.01674$	
Equatorial Radius of the Earth	$R_e = 6378.137$	km

---

*Reference: PHYSICS TODAY, July 2014, (physicstoday.org) David B. Newell, A More Fundamental International System of Units, pp. 35-41

www.ingramcontent.com/pod-product-compliance
Lightning Source LLC
Chambersburg PA
CBHW020733180526
45163CB00001B/218